THE ʻAM HA-ARETZ

ARBEITEN ZUR LITERATUR UND GESCHICHTE DES HELLENISTISCHEN JUDENTUMS

HERAUSGEGEBEN VON

K. H. RENGSTORF

IN VERBINDUNG MIT

G. DELLING, H. R. MOEHRING,
B. NOACK, H. M. ORLINSKY, H. RIESENFELD,
A. SCHALIT, H. SCHRECKENBERG, W. C. VAN UNNIK,
A. WIKGREN, A. S. VAN DER WOUDE

VIII

AHARON OPPENHEIMER

THE ʻAM HA-ARETZ

LEIDEN
E. J. BRILL
1977

THE ʿAM HA-ARETZ

A Study in the Social History of the Jewish People in the Hellenistic-Roman Period

BY

AHARON OPPENHEIMER

TRANSLATED FROM THE HEBREW BY I. H. LEVINE

LEIDEN
E. J. BRILL
1977

ISBN 90 04 04764 6

PRINTED IN THE NETHERLANDS

CONTENTS

FOREWORD

The idea of the present study was first put forward at a seminar held during the winter term of 1967. Mr. Oppenheimer had given a comprehensive lecture on the institution of the tithes in the period of the Second Temple, and in the lively discussion that followed the lecture I suggested to Mr. Oppenheimer that he takes the subject as a thesis and treats it in its wider context. Thus there was born the idea of setting the subject of the lecture within the broader framework of the concept of the ‘am ha-aretz in all its aspects, halakhic, sociological, and national.

In his extensive study Mr. Oppenheimer expresses first and foremost disagreement with the view of Zeitlin and of those who associate themselves with it, as well as disagreement with Büchler's theory, seeking as Mr. Oppenheimer does to liberate the concept of the ‘am ha-aretz from the Procrustean bed into which it was thrust by Büchler and to restore it to its rightful dimensions and historic proportions.

This is without doubt an historic work in the true meaning of the word, since its object is not to emphasize the fixed, static aspect of the subject but rather the process of its emergence in all its diversity. Not less important is the critical approach to the relation between the institution of the ‘am ha-aretz and Christianity. Mr. Oppenheimer establishes that there was no difference between the people of Galilee and those of Judaea as regards their attitude to the halakhah and its observance, in consequence of which there is no reason to view the growth of Christianity against the background of the emergence of a new doctrine for the masses, the ‘ammei ha-aretz, poor in a knowledge of the Torah and in the observance of the commandments.

It is to be hoped that the present work will stimulate a discussion of this important subject of the Second Temple period and of the first centuries after its destruction. So will this study best fulfil its object.

Jerusalem
1st August 1976

A. Schalit

PREFACE

Dealing as it does with the ʻam ha-aretz, the chief aim of this study is to elucidate the concept and to shed light on its social manifestations. The principle that modifications occur in the incidence of sociological concepts holds good also as far as the concept of the ʻam ha-aretz is concerned. Accordingly, this study seeks to ascertain the changes and developments that occurred in the social incidence of the concept of the ʻam ha-aretz.

The main chapters of the study are structured in the form of a spiral. Thus the chapter "Tithes and Ritual Purity" seeks to explain some of the concepts relating to the ʻam ha-aretz; the chapter "The ʻAm ha-Aretz le-Mitzvot and the ʻAm ha-Aretz la-Torah" deals largely with the halakhic status of the ʻam ha-aretz; the chapters "The ʻAmmei ha-Aretz, the Pharisees, and the Ḥaverim" and "The Relations between the ʻAmmei ha-Aretz and the Talmidei Ḥakhamim" treat the subject on the historical-social level. In general, the aim is to determine the social status of the ʻam ha-aretz and his place in the social context.

In talmudic literature the ʻam ha-aretz is contrasted with ḥaverim and with talmidei ḥakhamim. In the course of our treatment of the subject we have also dealt with these concepts but not exhaustively except in so far as they concern areas that relate to the ʻam ha-aretz.

The preparatory work for this study included an examination of epigraphic and papyrologic evidences to find out if manifestations analogous to the ʻam ha-aretz existed in the cultures of neighbouring peoples in the period under discussion. This research produced no results. It became evident that the ʻam ha-aretz in its halakhic and social aspects is an internal Jewish occurrence and hence no analogous manifestations were found. However, there is definitely a relation between, on the one hand, the associations of the ḥaverim and the other sects in Eretz Israel in the period of the Second Temple and of the Mishnah and, on the other hand, the organization into associations in the Greek-Hellenistic-Roman world. External institutions are to be seen as having had an influence on the emergence of the associations in Eretz Israel. This is a subject to which a special article should be devoted.

The last chapter of this study "The ʻAmmei ha-Aretz, the Christians, and the Samaritans'' is in the nature of an appendix which deals with the relation of the 'am ha-aretz to the other social-religious manifestations on the fringes of Jewish society in the period covered by this study. There is no intention here to identify the 'am ha-aretz with the Christian-Jewish community or with the Samaritans but rather to deal with the relations between them.

The study seeks to examine the place of the 'am ha-aretz in urban and rural settlements and in the various regions of Eretz Israel — Judaea, Galilee, and Transjordan. We have not been able to find an answer to the question what was the percentage of the 'ammei ha-aretz in the Jewish community and what was the extent of this manifestation. From the sources it is clear that their number was large, but we have been unsuccessful in our efforts to find a precise means of arriving at an answer to this question.

It should be added that all that we know about the nature of the 'ammei ha-aretz and about their approach to various subjects derives from the statements of the Sages. There is, of course, extant no ''Talmud of the 'Ammei ha-Aretz,'' so that in general we only know how the social elite looked on the 'ammei ha-aretz and are unable to learn from the 'ammei ha-aretz themselves about their views and attitudes.

The original Hebrew version of this study was submitted to the Senate of the Hebrew University of Jerusalem for the degree of Ph.D. When writing it, I had the benefit of the constant guidance and advice, criticism and encouragement of Professor Samuel Safrai and Professor Abraham Schalit.

To Dr Israel Levine, who has translated the work, I wish to express my appreciation.

Several Jerusalem scholars were consulted by me — Professor E. E. Urbach, Professor Ch. Rabin, and Professor S. E. Loewenstamm — to all of whom I express my thanks.

I am indebted to Mr M. Kosovsky for placing at my disposal the material for the concordance of the Jerusalem Talmud which he is compiling.

A. O.

Jerusalem
15th Shevat 5735
27th January 1975

ABBREVIATIONS

BOOKS OF THE BIBLE

Genesis	Gen.	Ezekiel	Ezek.
Exodus	Ex.	Hosea	Hos.
Leviticus	Lev.	Zephaniah	Zeph.
Numbers	Num.	Zechariah	Zech.
Deuteronomy	Deut.	Malachi	Mal.
Joshua	Jos.	Psalms	Ps.
Judges	Jud.	Ecclesiastes	Eccl.
Samuel	Sam.	Esther	Est.
Isaiah	Is.	Nehemiah	Neh.
Jeremiah	Jer.	Chronicles	Chron.

Maccabees .... Macc.

BOOKS, JOURNALS, ETC.

AJ	Josephus, Jewish Antiquities.
ASTI	Annual of the Swedish Theological Institute.
AV	Authorized Version of the Bible.
BCE	Before the Common Era.
BJ	Josephus, The Jewish War.
CE	Common Era.
CPJ	Corpus Papyrorum Judaicarum.
DJD	Discoveries in the Judaean Desert.
HTR	Harvard Theological Review.
IEJ	Israel Exploration Journal.
JBL	Journal of Biblical Literature.
JJLG	Jahrbuch der jüdisch-literarischen Gesellschaft.
JJS	Journal of Jewish Studies.
JNES	Journal of Near Eastern Studies.
JQR	Jewish Quarterly Review.
JSS	Journal of Semitic Studies.
JThS	Journal of Theological Studies.
MGWJ	Monatsschrift für Geschichte und Wissenschaft des Judentums.
OTS	Oudtestamentische Studien.
RA	Revue d'Assyriologie.
REJ	Revue des Études Juives.
RV	Revised Version of the Bible.
TB	Babylonian Talmud.
TJ	Jerusalem Talmud.
VT	Vetus Testamentum.
ZAW	Zeitschrift für die alttestamentliche Wissenschaft.
ZNW	Zeitschrift für die neutestamentliche Wissenschaft.

TRANSLITERATION OF HEBREW

The following method is used for the transliteration of Hebrew words and names (except for names which occur in the Bible, when the forms as they appear in English translations of the Bible are used).

א	Indicated by ' — except at beginning and end of word.
בּ	b
ב	v
גּ ג	g
דּ ד	d
ה	h
ו	v
ז	z
ח	ḥ
ט	ṭ
י	y
כּ	k
כ	kh
ל	l
מ	m
נ	n
ס	s
ע	ʻ
פּ	p
פ	f
צ	tz
ק	ḳ
ר	r
שׁ	sh
שׂ	s
תּ ת	t

a) Dagesh lene is not indicated, except in the letters בּ, כּ, and פּ. Dagesh forte is indicated by doubling the letter, except in the case of *sh* and of *tz*.

b) The Hebrew definite article is indicated by *ha* or *he* followed by a hyphen, but without the following letter doubled.

CHAPTER ONE

INTRODUCING THE SUBJECT

1. Trends in and Inquiries into the Subject

Talmudic literature makes frequent reference to the concept of the 'am ha-aretz, which thus occurs in sources that span various periods and range over a wide variety of topics. A synoptic glance at these sources reveals at the outset only this — that to apply to an individual or a group the term 'am ha-aretz is to criticize or disparage that individual or group.

The question "Who is an 'am ha-aretz?" is discussed by tannaitic sources which give various definitions of the concept, such as —

דתניא איזהו עם הארץ, — כל שאינו אוכל חוליו בטהרה, דברי רבי מאיר.
וחכמים אומרים, כל שאינו מעשר פירותיו כראוי...
תנו רבנן, איזהו עם הארץ, —
כל שאינו קורא קריאת שמע ערבית ושחרית, דברי רבי אליעזר.
רבי יהושע אומר, כל שאינו מניח תפילין.
בן עזאי אומר, כל שאין לו ציצית בבגדו.
רבי נתן אומר, כל שאין מזוזה על פתחו.
רבי נתן בר יוסף אומר, כל שיש לו בנים ואינו מגדלם לתלמוד תורה.
אחרים אומרים, אפילו קרא ושנה ולא שמש תלמידי חכמים הרי זה עם הארץ

"It has been taught, Who is an 'am ha-aretz? Anyone who does not eat his secular food in ritual purity. This is the opinion of R. Meir. The Sages however said, Anyone who does not tithe his produce properly... Our Rabbis taught, Who is an 'am ha-aretz? Anyone who does not recite the shema' evening and morning. This is the view of R. Eliezer. R. Joshua said, Anyone who does not put on tefillin. Ben 'Azzai said, Anyone who does not have tzitzit on his garment. R. Nathan said, Anyone who does not have a mezuzah at his door. R. Nathan b. Joseph said, Anyone who has sons and does not bring them up to a study of the Torah. Others again said, Anyone who has learnt Scripture and Mishnah but has not ministered to talmidei ḥakhamim is an 'am ha-aretz."[1]

---

[1] TB Berakhot 47b; and cf. TB Giṭṭin 61a; Tosefta 'Avodah Zarah iii, 10; TB Soṭah 21b–22a.

During the past hundred years or so, several historians and scholars of talmudic literature have sought to define the precise meaning of the concept of the 'am ha-aretz and its social incidence, its date and place in the historical situation, and its relation to spiritual, social, and other factors in the nation's history.

In the sources the 'am ha-aretz is frequently contrasted with the ḥaver, in that the latter was scrupulous in observing the commandments relating to the tithes and to ritual purity, whereas the former was not so scrupulous in these matters.[2] The 'am ha-aretz is also contrasted with the talmid ḥakham, for whereas the latter's interests centered in the Torah, the former was not only ignorant of but lacked any desire to study it.[3] That there was a seething mutual hatred between the talmid ḥakham and the 'am ha-aretz is attested by the sources.[4]

On the basis of some or all of these statements a considerable number of scholars, who dealt with the period of the Second Temple and of the Mishnah, concluded that in one way or another the term 'am ha-aretz is synonymous with an ignoramus and/or refers to one who was not scrupulous in observing all or some of the commandments.

Most of these scholars linked the 'am ha-aretz with the emergence of Christianity, since at its inception it was opposed, so runs one of their arguments, to some of the things that the 'ammei ha-aretz were said to disregard.[5] Viewing Christianity as emerging from the world of the 'ammei ha-aretz, these scholars saw in Jesus and his disciples the representatives of the 'ammei ha-aretz in their fight against the Pharisees and the Sages. The 'am ha-aretz was furthermore associated by some of these scholars with Galilee, the cradle of Christianity, where there were, so they contended, masses poor in the observance of the commandments and in the study of the Torah. To them Jesus came, proclaiming salvation.[6]

While it is generally agreed among scholars that the 'ammei ha-aretz

---

[2] See below, pp. 69–96.

[3] See below, pp. 97–106.

[4] See below, pp. 172–180.

[5] See below, pp. 218–229.

[6] E. Schürer, *Geschichte des jüdischen Volkes im Zeitalter Jesu Christi*, Leipzig, 1907[4], II, pp. 454–475; W. Bousset, *Die Religion des Judentums*, Tübingen, 1926[3], pp. 187 ff.; G. F. Moore, *Judaism in the First Centuries of the Christian Era*, Cambridge (Mass.), 1927–1930; M. Friedländer, *Zur Entstehungsgeschichte des Christentums* Vienna, 1894, pp. 37–58; *idem*, *Die religiösen Bewegungen*, Berlin, 1905, pp. 78–88; K. Kohler, *Jewish Encyclopedia*, I, pp. 484–485; J. Klausner, *Historyah shel ha-Bayit ha-Sheni*, Jerusalem, 1958[5], in particular IV, pp. 220 ff.; etc.

were the ordinary masses, some endeavoured to define their social class more precisely.

The ʻam ha-aretz is the subject of a special study by Zeitlin,[7] who sought to determine the social significance of the concept and to trace its development.

Maintaining that at first the ʻam ha-aretz denoted the farmer, the peasant, Zeitlin quotes in support of this contention the Baraita תניא רבן שמעון בן גמליאל אומר, הלכות הקדש תרומות ומעשרות, הן הן גופי תורה ונמסרו לעמי הארץ "It was taught, Rabban Simeon b. Gamaliel said, The halakhot of heḳdesh,[a] terumot, and tithes are essential parts of the Torah, and were entrusted to the ʻammei ha-aretz."[8] To the nature of their occupation and to their class Zeitlin assigns the reasons for their siding with the Pharisees against the Boethusians (Sadducees) that the commandment of the willow could be observed on the seventh day of Tabernacles which fell on a Sabbath.[9] Since on this festival judgment is passed in respect of water and since rain is the farmer's chief concern, it was clearly important for the farmer ʻam ha-aretz that, in particular, the commandments of Tabernacles should be fulfilled. There is, according to Zeitlin, no justification for regarding the ʻam ha-aretz as an ignoramus, since the Baraita which refers to בן חבר שלמד אצל עם הארץ "the son of a ḥaver who studied under an ʻam ha-aretz"[10] shows that the ʻammei ha-aretz were not ignorant people.

According to this view, the emergence and development of the concept of ʻam ha-aretz were as follows. After the Restoration the majority of the population consisted of the priestly and levitical class on the one hand and the farmer ʻam ha-aretz class on the other. At this juncture an inter-class confrontation had not as yet taken place, most of the people being in fact ʻammei ha-aretz. After the Ptolemaic conquest of the country, and particularly after the Hasmonaean revolt, the form of society in Eretz Israel changed. Cities on the coast were added to Judaea, trade and commerce began to flourish, and a new class of urban population emerged to which the obligations of the tithe did not apply, thus discriminating against the farmer ʻam ha-aretz as compared

---

7 S. Zeitlin, "The Am haarez," *JQR* (N.S.), 23 (1932), pp. 45–61.

8 TB Shabbat 32a–b. This Baraita also occurs in Tosefta Shabbat ii, 10.

9 Tosefta Sukkah iii, 1; TB Sukkah 43b.

10 Tosefta Demai ii, 18. There is no proof that the reference here is to an ʻam ha-aretz who taught Torah to a ḥaver's son. It more probably refers to the teaching of a handicraft.

a) An object consecrated to the Temple.

with the urban merchant. Thereupon the ʻam ha-aretz stopped separating tithes. At this stage a class conflict was engendered and produced the first laws relating to the ʻam ha-aretz,[11] among them being the obligation of anyone purchasing produce from the ʻam ha-aretz to separate the tithes.

The most serious breach occurred between, on the one hand, the urban leaders, to whom Zeitlin applies the term ḥaverim, and who constituted the privileged aristocracy, and, on the other, the ʻammei ha-aretz, the plebeian tillers of the soil.

Zeitlin also contends that among those who joined Christianity were many of the ʻammei ha-aretz.[12] Confirmation of this he finds in the verse, "Have any of the authorities or of the Pharisees believed in him? But this crowd, who do not know the law, are accursed."[13] In the tannaitic literature of the first century the term ʻam ha-aretz frequently refers to Jewish-Christians, for the ʻam ha-aretz was defined as one who did not observe the precepts of the tefillin and of the tzitzit, the very commandments which were opposed by the Jewish-Christians, as can be seen from the verse in Matthew, "They do all their deeds to be seen by men; for they make their phylacteries broad and their fringes long."[14] Thus the ʻammei ha-aretz and the Jewish-Christians were identical.

In later tannaitic literature the term ʻam ha-aretz acquired a third meaning, when it referred not only to farmers and to Jewish-Christians, but also became a general designation for all whom the Sages considered crude, immoral, and lawless.

Such are Zeitlin's views, which have been followed in part or in full by other scholars.[15]

The most extensive study on the subject of the ʻam ha-aretz and one that uses the largest number of sources is Büchler's work on the Galilaean ʻam ha-aretz in the second century.[16]

---

[11] These were, according to Zeitlin, the ordinances of Johanan the high priest. For these ordinances, see below, pp. 34–36.

[12] We shall not give here Zeitlin's views on the relation of the ʻam ha-aretz to the Great Revolt against Rome or to the political and spiritual trends preceding it.

[13] John 7:48–49.

[14] Matthew 23:5; cf. Justin, *Dialogue with Tryphon*, 46.

[15] The ʻam ha-aretz has been dealt with in a manner similar to that of Zeitlin by S. W. Baron, *A Social and Religious History of the Jews*, New York, 1952², II, pp. 67–68; and by L. Finkelstein, *The Pharisees*, Philadelphia, 1962³. Reference is made to these views on pp. 18 ff. and *passim*, below.

[16] A. Büchler, *Der galiläische ʻAm-haʼAreṣ des zweiten Jahrhunderts*, Vienna, 1906.

The basis of Büchler's approach is that a considerable number of the halakhot and statements dealing with the ‘am ha-aretz occurs in the names of Tannaim of the school of Usha, among these being R. Meir, R. Judah b. Ila‘i, R. Jose b. Ḥalafta, R. Nehemiah, Rabban Simeon b. Gamaliel, and R. Simeon b. Yoḥai.

Accordingly, Büchler maintained that the ‘am ha-aretz le-mitzvot, who was suspected of disregarding the commandments associated with the produce of the land in Eretz Israel [מצוות התלויות בארץ] and the precepts of ritual purity, was a social-religious concept which referred to the period of Usha and to Galilee.

Any reference to the ‘am ha-aretz who belonged neither to the period of Usha nor to Galilee was to the ‘am ha-aretz la-Torah, that is, one ignorant of the Torah.

Thus, according to Büchler, there were two different concepts of the ‘am ha-aretz — the general concept of the ‘am ha-aretz la-Torah which was used at various periods and which applied also to Judaea, and the concept of the ‘am ha-aretz le-mitzvot, which referred exclusively to the period of Usha and was restricted to Galilee.

It was to the ‘am ha-aretz le-mitzvot that Büchler related the expressions of hatred of the ‘am ha-aretz, since he regarded such expressions as referring principally to the period of Usha and to Galilee.

The emergence of the concept of the ‘am ha-aretz le-mitzvot in the period of Usha was explained by Büchler as due to historical circumstances. After the Bar Kokheva revolt, the leading Sages came to Galilee, and there re-established the leadership of the nation. Finding Galilee empty of a knowledge of the Torah and devoid of the observance of the commandments, they applied the disparaging term of ‘am ha-aretz to the Galilaeans in an attempt to goad them into observing the commandments and keeping the Torah.

Not satisfied with forcing the concept of the ‘am ha-aretz le-mitzvot into this Procrustean bed of period and locality, Büchler went on to maintain that most of the halakhot relating to the ‘am ha-aretz and dealing with ritual purity and with the eating of secular food in a state of ritual purity were addressed in the main to the priests, who were commanded to observe the ritual purity of terumah. According to Büchler, these halakhot did not, then, apply to the people as a whole, but to the priestly ḥaverim and the priestly ‘ammei ha-aretz, the former of whom were, and the latter were not, scrupulous in observing the halakhot of ritual purity and impurity.

Without going, for the moment, into details, we wish to make several general comments on Büchler's views. Some of these remarks are cited from the writings of G. Allon and E. E. Urbach.[17]

To begin with, it should be pointed out that Büchler's approach displays a certain tendency towards apologetics. In his day, Christian scholars, greatly emphasizing the opposition that existed between the ḥaver and the 'am ha-aretz, sought, as mentioned above, to explain the activities of Jesus as reformative in character, in that Jesus in his sermons advocated the cause of the 'am ha-aretz and denounced the gap prevailing between the Pharisees and the ḥaverim on the one hand and the common people on the other.

By setting out, as he did, to prove the lateness not only of all restrictive halakhot which prohibited a close association between ḥaverim and 'ammei ha-aretz, but also of the expressions indicative of a hatred between the latter and talmidei ḥakhamim, Büchler deprived this theory of the Christian scholars of any actual, historical basis, for according to him these halakhot originated under the conditions of Galilee at a late period and hence had nothing to do with Jesus and with his activities.

As we have seen, Büchler's main argument is that those who transmitted the halakhot relating to the 'am ha-aretz le-mitzvot were generally Sages of the generation of Usha. But it is not solely on the basis of its author that the date of an halakhah is to be fixed. For the halakhah has to be examined in respect of its contents and context, as well as of its place within the prevailing conditions, since it is generally agreed that later Sages transmitted statements which belonged to earlier periods and Sages.

As for the halakhot transmitted in the days of Usha, a very thorough investigation has to be made into them. The Sages of that period were engaged in compiling the tannaitic statements and in laying the basis of the close of the Mishnah under R. Judah ha-Nasi. There was almost no halakhic subject that was not dealt with in the schools of the Sages of Usha. It is clearly impossible to determine, precisely for the period of Usha, the historical setting of concepts and of halakhot solely on the basis of the fact that the names associated with them are those of the Sages of Usha.

---

[17] G. Allon, *Toledot ha-Yehudim be-Eretz Yisra'el bi-Teḳufat ha-Mishnah ve-ha-Talmud,* Tel Aviv, I: 1958³, II: 1961² — see especially I, pp. 160–163, 315–319, II, pp. 80–83, 135–136; *idem, Meḥḳarim be-Toledot Yisra'el,* Tel Aviv, 1957–1958, and see especially I, pp. 148–176, II, pp. 58–73. See also the writings of E. E. Urbach, in particular his *Ḥazal — Pirḳei Emunot ve-De'ot,* Jerusalem, 1969, pp. 520–580.

Allon[18] states that it is difficult to grasp what, according to Büchler's contention, could have prompted the Sages to pass restrictive laws and introduce new prohibitions relating to ritual purity a considerable time after the destruction of the Second Temple, when any direct association between the people and the Temple and its sacred things had long ceased.[19]

It is on the Galilaeans' sparse knowledge of the Torah and disregard of the commandments that Büchler based his view that the halakhot pertaining to the 'am ha-aretz were instituted specifically for Galilee. This view, which sets the world of the Sages and of the Torah, until the Bar Kokheva revolt, in Judaea and leaves Galilee peopled by the 'ammei ha-aretz, is not, as we have seen, limited to Büchler. Many scholars locate the 'ammei ha-aretz in Galilee, precisely in the days of the Second Temple and in the period of Jabneh. But this view has no foundation in fact. For life in Galilee bore the halakhic-Pharisaic stamp exactly as in Judaea, and no distinction can be made between the Galilaean and the Judaean Jew. Nor can it be said that there was a special Galilaean type or a different Galilaean pattern of life.[20]

There are moreover halakhot relating to the 'am ha-aretz le-mitzvot which originated in the days of the Second Temple, and Büchler's attempt to assign them to a later date cannot be sustained.[21] In general it may be said that in order to prove his thesis Büchler often did violence to texts.

We shall give one example of how, on the subject of eating secular food in a state of ritual purity, Büchler analysed texts to fit in with his views.

Büchler limited ritual purity to the Temple and to the priests. The halakhot, which were restrictive in the sphere of ritual purity and impurity and which prohibited contact with the 'ammei ha-aretz, originated, according to him, after the destruction of the Second Temple, some in the days of Jabneh but the vast majority in the generation of Usha. Even these halakhot were primarily intended for the priests and applied

---

[18] See Allon, *op. cit.*, *loc. cit.*

[19] It can however be contended that ritual purity was practised, perhaps even more intensely, also after the destruction of the Second Temple, although Allon is undoubtedly right in stating that it is difficult to assume that new prohibitions were instituted specifically in the days following the destruction of the Temple. See below, pp. 63–66.

[20] See below, pp. 200–217.

[21] For details, see below, pp. 69–96.

to the terumah which priests were obliged to eat in a state of ritual purity, although they sometimes observed the same standard of ritual purity also as regards their secular food.[22] To prove his point of view Büchler used the sources as follows.

a) He maintained that ancient traditions were formulated by the Sages of Usha.[23]

In some aspects of the halakhah dealing with ḥaverim there was a difference of opinion between Bet Hillel and Bet Shammai. עד מתי מקבלין, בית שמאי אומרים למשקין שלשים יום ולכסות שנים עשר חודש, ובית הילל אומרים זה וזה שלשים יום "How long is the period that has to elapse before he is accepted as a ḥaver? Bet Shammai said, As regards the purity of liquids, the period is thirty days, and as regards the purity of his garments, the period is twelve months, whereas Bet Hillel said, In both cases the period is thirty days."[24] Contending that this halakhah was formulated in the days of Usha, Büchler based his view on the fact that in the chapter of the Tosefta in which it occurs, this particular halakhah is included in a collection of halakhot of the generation of Usha.

b) The scope of undefined traditions was limited by Büchler in accordance with the other traditions in that context.[25]

In Mishnah Ṭohorot we learn המפקיד כלים אצל עם הארץ, טמאים טמא מת וטמאין מדרס. אם מכירו שהוא אוכל בתרומה, טהורין מטמא מת אבל טמאין מדרס "If a man deposited vessels with an 'am ha-aretz they are deemed to be impure with corpse-impurity and with midras-impurity.[a)] If he knows him to eat terumah, they are free from corpse-impurity but are impure with midras-impurity."[26] This Mishnah refers to an 'am ha-aretz who is not a priest, for "if he knows him to eat terumah" merely states a supposition. Nevertheless Büchler contended that since

---

[22] Büchler, *op. cit.*, *passim*, and see, in particular, pp. 64–96. The general demand to eat secular food in ritual purity was regarded by Büchler as an ethical and educational precept but not as a decided law. We shall later revert to the subject of the eating of secular food in ritual purity; and see especially pp. 118–156, below.

[23] Büchler, *op. cit.*, p. 170, note 2.

[24] Tosefta Demai ii, 12; and cf. TB Bekhorot 30b; these halakhot are discussed on pp. 121–129, below.

[25] Büchler, *op. cit.*, p. 49, and see also pp. 83–85.

[26] Ṭohorot viii, 2. This halakhah and similar ones are discussed below, on pp. 164–169.

a) Impurity contracted by an object on which one with a discharge (see Lev. 12:2; 15:2, 25) sits, treads, lies, or leans. The impurity was conveyed to anyone who, in this case, carried, or was carried on, the impure vessels deposited.

this Mishnah occurs in a context of Mishnayot that deal with priests and with terumah, it too is of the same type, and the supposition "if he knows him to eat terumah" shows that there were priests who did not eat terumah, and it is accordingly to them that the first part of the Mishnah also refers.[27]

c) Büchler distorted the meaning.[28]

A Tosefta which sets forth the obligations of the ḥaver [in contrast to the 'am ha-aretz] states, **ושיהא אוכל חוליו בטהרה** "And he eats his secular food in ritual purity."[29] There is apparently no hint here that this refers to a priest. But Büchler, straining the fact that the passage mentions "his secular food," contended that it related to one who also had food that was not secular, that is, a priest who had terumah in his possession.

This is a case of pressing the meaning of the passage too far, so that its sense is distorted. What is more, other MSS. of the Tosefta, as well as parallel passages, have the reading, "And he eats secular food [not 'his secular food'] in ritual purity."[30]

The actual, historical background of the halakhot which contrast the 'am ha-aretz le-mitzvot with the ḥaver has become clearer through the Dead Sea Scrolls, in particular the *Manual of Discipline*, which deals with the laws of the community of the Yaḥad. There is undoubtedly a great similarity between the laws relating to the ḥaver and those pertaining to the Yaḥad,[31] and the ḥaverim and the Yaḥad have, it seems, to be assigned to the same period. We adopt the view of those scholars who hold that the Dead Sea sects are to be referred to the last days of the Second Temple, which is further proof of the early date of the 'am ha-aretz le-mitzvot, thus demolishing Büchler's theory.

The most recent scholar to deal in detail with the question of the 'am ha-aretz has been E. E. Urbach in his comprehensive work *Ḥazal* —

---

27 In no other source does a reference to those who eat terumah apply to only some priests. This Mishnah can also be interpreted as referring to a priest who deposits vessels with the 'am ha-aretz, who knows him [i.e. the priest] to eat terumah. The 'am ha-aretz can in every case be a layman.

28 Büchler, *op. cit.*, p. 161.

29 Tosefta Demai ii, 2.

30 This is the reading in the Vienna MS. and in printed editions, in Avot de-R. Nathan, Version A, xli, ed. Schechter, p. 132, in the Munich MS. of TB Berakhot 47b, and in *Kaftor va-Peraḥ*, xxxvi, ed. Edelmann, 83b, on Tosefta 'Avodah Zarah iii, 10.

31 See below, pp. 147–151.

Pirḳei Emunot ve-De'ot.[32] While accepting in principle the distinction made by Büchler between the 'am ha-aretz le-mitzvot and the 'am ha-aretz la-Torah, Urbach basically rejects Büchler's other views on the 'am ha-aretz mentioned above. Thus Urbach links the 'am ha-aretz le-mitzvot and the ḥaverim with the sect of the Yaḥad. He endeavours to fix the date when there came into use the concepts of the 'am ha-aretz le-mitzvot and the 'am ha-aretz la-Torah, the former of whom was contrasted with the ḥaver in the association who was scrupulous in separating tithes and in observing ritual purity, while the latter was contrasted with the ḥaver in the sense of a talmid ḥakham. Urbach also discusses the acrimonious statements made by the Sages in denouncing the 'am ha-aretz.[33]

2. The Emergence and Development of the Concept of the 'Am ha-Aretz

The term 'am ha-aretz occurs first in the Bible. Except perhaps in those biblical books that deal with the beginning of the Second Temple period, the term in the Bible has, it seems, no connection with the one used from the rise of the Hasmonaean kingdom onwards.

When it occurs in the Bible in the singular, the term 'am ha-aretz, which is never used in connection with the kingdom of Ephraim, refers generally to a special social unit in the kingdom of Judah. What the precise nature of this social unit was biblical scholars have so far been unable to define. However the 'am ha-aretz occurs alongside other social strata which constituted part of the nation's leadership, such as princes, prophets, and priests,[34] and among the activities of the 'am ha-aretz mentioned in the Bible was participation in the revolt against Athaliah[35] and in proclaiming kings.[36] All this attests to the high social status of the 'am ha-aretz, but whether the term denotes a defined institution with specific functions, or a political, economic, and social stratum is the subject of divergent opinions among biblical scholars.[37]

---

32 See note 17, above.

33 See *op. cit.*, chap. 16.

34 Jer. 1:18; 34:19; 37:2; Ezek. 7:27; 22:29.

35 II Kings 11.

36 II Kings *loc. cit.*; 23:30.

37 See the literature on the 'am ha-aretz in the Bible: Ch. Tchernowitz, *Toledot ha-Halakhah*, New York, 1935–1950, III, pp. 19 ff.; E. Urbach, " 'Am ha-Aretz,"

When the term 'am ha-aretz occurs in the Bible in the plural, in the form of 'ammei ha-aretz or 'ammei ha-aratzot, it generally denotes foreigners, in contrast to the children of Israel, and it can then mean either the peoples of the world or those living in Eretz Israel.[38] In one case, that of the purchase of the cave of Machpelah, the term signifies not the inhabitants of a certain place but a group of men with a special, defined function.[39]

In the biblical books which describe the period of the Restoration and the early days of the Second Temple, the term 'ammei ha-aretz is contrasted with the returned exiles,[40] who, so the Bible states, were those who returned to Zion after having been purified and after having repented in the captivity. Scholars however differ whether the 'ammei ha-aretz here refer to the people of Judah who remained in the country after the destruction of the First Temple and adopted a syncretistic faith, or whether in this instance too foreigners are meant.[41] A difference is made between 'ammei ha-aretz and those "who have separated themselves from the 'ammei ha-aratzot."[42] Thus as early as at the beginning of the Second Temple period a distinction was made in Jewish society between the loyal elements and those close to foreigners or to

---

Proceedings of the World Congress of Jewish Studies, 1947, Jerusalem, 1952, pp. 362–366; S. Zemirin, *Yo'shiyyahu u-Teḳufato*, Jerusalem, 1952, pp. 66–87; Y. Kaufmann, *Toledot ha-Emunah ha-Yisra'elit*, Jerusalem and Tel Aviv, 1936–1956, IV, 1, pp. 183–184; S. Talmon, *Bet Miḳra*, 31 (1967), pp. 27–55; M. Sulzberger, *The Am ha-aretz, The Ancient Hebrew Parliament*, Philadelphia, 1910²; E. Klamroth, *Die jüdischen Exulanten in Babylonien*, Leipzig, 1912, pp. 99–101; N. Slousch, *JQR* (N.S.), 4 (1913–1914), pp. 303–310; J. Wellhausen, *Israelitische und jüdische Geschichte*, Berlin, 1958⁹, p. 154; M. Weber, *Das antike Judentum*, Tübingen, 1921, pp. 30 ff.; E. Gillischewski, *ZAW*, 40 (1922), pp. 137–142; S. Daiches, *JThS*, 30 (1929), pp. 245–249; L. Rost, *Festschrift-Procksch*, Leipzig, 1934, pp. 125–148; R. Grodis, *JQR* (N.S.), 25 (1934–1935), pp. 237–259; E. Würthwein, *Der 'amm ha'arez im AT*, Stuttgart, 1936; I. L. Seeligmann, *The Septuagint Version of Isaiah*, Leyden, 1948; C. U. Wolff, *JNES*, 6 (1947), pp. 98–108; G. von Rad, *Deuteronomiumstudien*, Göttingen, 1948²; I. D. Amussin, *Vestnik Drevney Istorii*, 2 (1955), pp. 14–36; R. de Vaux, *Institutions*, Paris, 1958-1960, I, pp. 111–113; *idem*, *RA*, 58 (1964), pp. 167–172; J. A. Soggin, *VT*, 13 (1963), pp. 187 ff.; E. W. Nicholson, *JSS*, 10 (1965), pp. 59–66; S. Talmon, *Proceedings of the Fourth World Congress of Jewish Studies, 1965*, Jerusalem, 1967, I, pp. 71–76; A. Malamat, *IEJ*, 18 (1968), p. 140; H. Tadmor, *Journal of World History*, 11 (1968), pp. 56–68.

38 Deut. 28:10; Jos. 4:24; I Kings 8; Zeph. 3:20; etc.

39 Gen. 23.

40 Neh. 10:29; and cf. Ezra 9:1; 6:19–21; etc.

41 See below, pp. 83–84.

42 Neh. 10:29.

assimilationists, who are here mentioned in relation to the ‘ammei ha-aretz.

From the rise of the Hasmonaean kingdom onwards to the end of the Second Temple period, as well as during the days of Jabneh and of Usha, as also in those of R. Judah ha-Nasi, and until the beginning of the period of the Eretz Israel Amoraim, the term ‘am ha-aretz signified alike an active social concept and an element in the social stratification of the people of Israel.

It is mainly in terms of two notions that the ‘am ha-aretz has been defined. The one is the ‘am ha-aretz le-mitzvot, the ‘am ha-aretz being disparaged for not scrupulously observing either the commandments associated with the produce of the land in Eretz Israel or the precepts relating to ritual purity. The second main notion in the definition of the term is the ‘am ha-aretz la-Torah, the ‘am ha-aretz being stigmatized as an ignoramus who has not studied the Torah. In both instances the term ‘am ha-aretz is used as a derogatory designation of one at the bottom of the social ladder, with the first type contrasted with the ḥaver who was scrupulous in observing the commandments disregarded by the ‘am ha-aretz, and the second type contrasted with the talmid ḥakham whose interests lay in the study of the Torah.[43]

At the centre of the Jewish world during most of the periods under discussion were the commandments associated with the produce of the land in Eretz Israel [in particular those relating to tithes and to the Sabbatical Year], the precepts pertaining to ritual purity, and the study of the Torah.

The separation of the tithes is a commandment whose dominant place was much emphasized and whose importance was greatly stressed.

In all generations the Sages were unsparing in urging and encouraging the people to separate the tithes. Numerous Baraitot and statements emphasized that the tithes were given in exchange for the gifts of Heaven, a notion somewhat similar to that embodied in the commandments of the first fruits and the first-born animals. Thus, for example, תני בשם רבי נחמיה, בנוהג שבעולם אדם יש לו שדה, והוא נותנה למחצה ולשליש ולרביע, אבל הקדוש ברוך הוא אינו כן, הק׳ משיב רוחות, ומעלה עננים, ומוריד גשמים, ומפריח טללים, ומגדל צמחים, ומדשן פירות, ולא אמר לנו להפריש אלא אחד מעשרה. לפיכך משה מזהיר את ישראל ואומר להם ״עשר תעשר״ “It was taught in the name of R. Nehemiah, A man who has a field generally hires it out for a share of a half, a third, or a quar-

[43] See below, pp. 67–117.

ter of its produce. But the Holy One, blessed be he, is not so. He causes the wind to blow, the clouds to rise, and the rain to fall, renders dew productive, makes plants grow and fruits juicy, and yet enjoins us to separate only one tenth. Accordingly Moses admonished Israel, saying to them, ‘You shall tithe’.”[44] Anyone who did not tithe was warned that his punishment by Heaven would be measure for measure, and the produce of his fields would suffer: רבי שמעון בן אלעזר אומר, הטהרה נטלה את הטעם ואת הריח, המעשרות נטלו את שמן הדגן “R. Simeon b. Eleazar said, [The cessation of] purity has deprived [fruit of its] taste and fragrance; [the cessation of] the tithes has deprived corn of its fatness.”[45] The Sages went even further. The failure to separate tithes would, they declared, be punished not only by a curse on the produce of the field but also by grave punishment from Heaven. תנו רבנן, אסכרה באה לעולם על המעשר “Our Rabbis taught, Croup comes to the world on account of [the neglect of] tithes.”[46] Or, אליהו זל שאל לרבי נהוריי, מפני מה באין זועות לעולם, אמר ליה בעון תרומה ומעשרות “Elijah asked R. Nehorai, Why do earthquakes come to the world? He answered him, Because of the sin of [the neglect of] terumah and tithes.”[47]

In contrast to the punishment in store for anyone who was not scrupulous in separating the tithes, praise and commendation were bestowed on those who separated their tithes properly. For example, בעא מיניה רבי מרבי ישמעאל ברבי יוסי, עשירים שבארץ ישראל במה הן זוכין, אמר ליה בשביל שמעשרין, שנאמר, ״עשר תעשר״ — עשר בשביל שתתעשר “Rabbi [R. Judah ha-Nasi] asked R. Ishmael b. Jose, Whereby do the wealthy in Eretz Israel merit [their wealth]? He replied, Because they give tithes, as it is written ‘ ‘asser te‘asser,’ [which means], give tithes [‘asser] that you may become wealthy [tit‘asher].”[48] The Midrash declares that God blessed the patriarchs because they gave tithes.[49] One who was scrupulous in giving tithes was referred to as a ne’eman, as a trustworthy person.[50] הוא [ר׳ ראובן בן אצטרובלי] היה אומר, ג׳ נאמנים הן, עני שהוא נאמן על הפקדון אין נאמן גדול מזה, בעל הבית שהוא נאמן על

---

44 Pesiḳta de-R. Kahana vi, 8, ed. Mandelbaum, p. 171.

45 Soṭah ix, 13; and cf. Avot de-R. Nathan, Version A, xxxviii, ed. Schechter, p. 111, and see the whole passage; cf. Version B, xli, ed. Schechter, p. 111.

46 TB Shabbat 33a–b.

47 TJ Berakhot ix, 13c.

48 TB Shabbat 119a; and similarly, TB Ta‘anit 9a.

49 Pesiḳta Rabbati xxv, ed. Friedmann, 127b.

50 For an explanation of the concept of a ne’eman, see below, pp. 151–156.

מעשרותיו אין נאמן גדול מזה, רווק שגדל במדינה ולא נחשד על הערוה אין נאמן גדול מזה "He [R. Reuben b. Estrobile (Strobilus)] said, There are three who are ne'emanim. A poor man who is trustworthy with regard to a deposit: there is no greater ne'eman than he. A householder who is trustworthy with regard to his tithes: there is no greater ne'eman than he. A bachelor who grew up in a large city and is not suspected of licentiousness: there is no greater ne'eman than he."[51] There is similarly the statement אמר רבי יוחנן, שלשה מכריז עליהן הקדוש ברוך הוא בכל יום, על רווק הדר בכרך ואינו חוטא, ועל עני המחזיר אבידה לבעליה, ועל עשיר המעשר פירותיו בצינעה "R. Johanan said, Concerning three does the Holy One, blessed be he, make proclamation every day, concerning a bachelor who lives in a large city without sinning, a poor man who returns lost property to its owner, and a wealthy man who tithes his produce unostentatiously."[52]

Although a large part of the above statements as also of the halakhic sources that deal with tithes date from the period after the destruction of the Second Temple, there are nevertheless aspects which reflect the attitude to tithes prevalent before its destruction. The compilation and redaction of the halakhah in the generations of Jabneh and of Usha and in the days of R. Judah ha-Nasi, the custom adopted by later Sages of quoting in their own names alike the statements of earlier Sages and things that mirror a prior historical situation, the attempt to continue the practice of the tithes also after the Second Temple period, all this led to the circumstance, mentioned previously, that most of the relevant statements and halakhot are from the days after the destruction of the Second Temple. An examination of the historical situation with the help both of external sources and of some of the statements and halakhot which can be shown to derive from the Second Temple period indicates that, as stated, the material referred to above reflects the situation which obtained during the existence of the Second Temple.

The preoccupation with the precept of the tithes in statements and in Midrashim, either praising and promising a reward to those who scrupulously separate their tithes or threatening those who fail to do so, attests not only to the importance with which the Sages viewed this

---

51 Avot de-R. Nathan, Version B, xxxv, ed. Schechter, p. 78.

52 TB Pesaḥim 113a. The Munich MS., Aggadot ha-Talmud, etc., do not read אמר רבי יוחנן "R. Johanan said," while the Munich B MS. has תנו רבנן "Our Rabbis taught."

precept but also to the attempt on the part of large numbers in every period to evade separating tithes as prescribed, for were this not so there would have been no need to reiterate praise on the one hand and threats on the other.[53]

It is therefore not surprising that the Sages should have chosen, as one of the descriptions with which to define the members of the reprehensible class of 'ammei ha-aretz, their disregard of the precept to separate tithes.

Besides the many sources which refer to him as not scrupulous in separating tithes, the 'am ha-aretz is also defined in individual halakhot and statements as not particular about the other commandments associated with the produce of the land in Eretz Israel, especially the commandment of the Sabbatical Year.[54]

There is an obvious economic difficulty in fulfilling this commandment, especially since to the biblical laws relating to the Sabbatical Year the Sages added further prohibitions.[55] Seeing as they did in the commandment of the Sabbatical Year a central injunction, the Sages were even stricter about it than about other commandments, as, for example, הנוטע בשבת, שוגג יקים, מזיד יעקר, ובשביעית, בין שוגג בין מזיד יעקר "He who plants anything on the Sabbath in error can allow it to remain, but if deliberately he must uproot it. But during the Sabbatical Year, whether [it was planted] unwittingly or deliberately he must uproot it."[56]

R. Judah ha-Nasi, who strove for the normalization of the economy and sought to introduce greater leniency into the laws of the Sabbatical Year, failed to gain the support of many contemporary Sages. Nor was his authority such as to enable him to abrogate these laws. רבי בעא מישרי שמיטתא, סלק רבי פנחס בן יאיר לגביה, א״ל מה עיבוריא עבידין, אמר ליה עולשין יפות, מה עיבוריא עבידין, א״ל עולשין יפות, וידע רבי דלית הוא מסכמא עימיה "Rabbi [R. Judah ha-Nasi] wished to permit [what was prohibited in] the Sabbatical Year. R. Phinehas b. Jair

[53] Since Philo (*De Specialibus Legibus*, I, §§153–154) strongly urged those neglectful of the tithes to fulfil the commandment, we infer that it was necessary also in his generation to induce people to tithe their produce.

[54] See below, pp. 67–117, and, in particular, pp. 80–82.

[55] See, for example, Sifra, Be-Har, iii, ed. Weiss, 108a; Shevi'it ii, 3; etc. On the Sabbatical Year, see S. Safrai, "Sabbatical Year Commandments under the Conditions Prevailing after the Destruction of the Second Temple" (Hebrew), *Tarbiz*, 35 (1966), pp. 304–328; 36 (1967), pp. 26–46.

[56] Terumot ii, 3; and cf. Tosefta Shabbat ii, 21; TB Giṭṭin 53b.

came to him. He [Rabbi] said to him, How are the crops? He answered, The endives are fine. How are the crops? He answered, The endives are fine. Thereupon Rabbi knew that he [R. Phinehas b. Jair] did not share his opinion.''[57]

The Sages viewed with disfavour the non-observance of the commandment of the Sabbatical Year even in times of religious persecution or of economic crisis. ״ביני ובין בני ישראל... אות היא לעולם״ — מגיד שאין השבת בטלה מישראל לעולם, וכן את מוצא, שכל דבר ודבר שנתנו ישראל נפשן עליהן נתקיימו בידן, וכל דבר ודבר שלא נתנו ישראל נפשן עליהן לא נתקיימו בידן — כגון השבת והמילה ותלמוד תורה וטבילה שנתנו נפשן עליהן נתקיימו בידן, וכגון בית המקדש והדינין ושמיטין ויובלות שלא נתנו ישראל נפשן עליהן לא נתקיימו בידן ''‘It is a sign for ever between me and the people of Israel.’ This asserts that the Sabbath will never disappear in Israel. Thus you find that anything on behalf of which Jews gave their lives has been preserved by them, and anything on behalf of which Jews did not jeopardize their lives has not been preserved by them. So, for example, the Sabbath and circumcision, the study of the Torah and immersion in a ritual bath, for which Jews gave their lives, have been maintained by them, but the Temple, civil courts, the Sabbatical and Jubilee Years, on behalf of which Jews did not imperil their lives, have not been maintained by them.''[58]

Here the Sabbatical Year is, it seems, put on the same level as the commandments for which a Jew יהרג ובל יעבור ''is to be killed rather than transgress them,'' and its importance is regarded as identical with that of the Temple itself. This statement also attests to an erosion in the observance of the Sabbatical Year among various sections of the people.[59]

---

[57] TJ Demai i, 22a; and cf. TJ Ta‘anit iii, 66b–c.

[58] Mekhilta de-R. Ishmael, Ki Tissa, i, ed. Horovitz-Rabin, p. 343.

[59] Interestingly enough, even at the beginning of the period of the Amoraim we find non-Jews at Caesarea ridiculing the Jews who observed the Sabbatical Year. רבי אבהו פתח, ״ישיחו בי יושבי שער,״ אלו אומות העולם שהן יושבין בבתי תרטיאות ובבתי קרקסיאות, ״ונגינות שותי שכר״ מאחר שהן יושבין ואוכלין ושותין ומשתכרין, הן יושבין ומשיחין בי ומלעיגים בי ואומרים, בגין דלא נצרוך לחרובא כיהודאי, והן אומרין אלו לאלו כמה שנים את בעי מחי, והן אומרים, כחלוקא דיהודאי דשבתא, ומכניסין את הגמל לטרטיאות שלהם והחלוקים שלו עליו והן אומרין אלו לאלו, על מה זה מתאבל, והן אומרים, היהודים הללו שומרי שביעית הן ואין להם ירק, ואכלו החוחים של זה והוא מתאבל עליהם ''R. Abbahu took up the text, ‘They that sit in the gate talk of me.’ These are the gentiles, who sit in the theatres and circuses. ‘And I am the song of the drunkards.’ After they sit and eat and drink and become drunk, they sit and talk of me, and ridicule me and say, We have no need to eat carobs like the Jews! And

Most of the halakhot and statements relating to the ʿam ha-aretz are concerned with the subject of ritual purity and impurity.

Ritual impurity is not viewed by Judaism as something perilous and frightening, but may rather be defined as a condition of negative taboo, transferable from one object to another. During the existence of the Second Temple and even after its destruction, the Sages made the laws of ritual purity a central principle within the totality of the commandments, heaped prohibitions on them, and dealt with them in innumerable halakhot. But it was particularly among sects that arose in Jewry in the days of the Second Temple, such as the Essenes and the Dead Sea sect, as well as among the associations of ḥaverim, that the ideal of ritual purity was made the dominant factor. The great strictness characterizing matters of ritual purity and impurity, the difficulty of complying with it, the danger of transferring ritual impurity from one person or object to another, all this led to a situation whereby ritual purity became the guiding principle in the division of Jewish society into classes. For the extremist sects, the stringent, scrupulous observance of ritual purity was one, if not *the*, reason for quitting the normative community. But also sects on its fringes, such as ḥaverim, as well as Pharisees who were a part of society, ascribed supreme value to the observance of ritual purity. Accordingly, whoever was not scrupulously observant of these commandments was defined as an ʿam ha-aretz and assigned to the lowest social class, while anyone more scrupulous in observing ritual purity was enjoined to exercise care not to come into contact with him.

Yet another central feature of Judaism in the tannaitic period was the study of the Torah, the significance of which increased particularly after the destruction of the Second Temple, as it progressively replaced the Temple service. The mission and aim now set before the Jew lay in the domain of the study of the Torah. This produced the term ʿam ha-aretz la-Torah, and also the disdain in which he was held for not becoming part of this stream.

An additional factor in the process of enhancing the value of the study

---

they say to one another, How long do you want to live? And they answer, As long as the Jew's Sabbath shirt [which is transmitted from father to son]. They bring a camel with its covers on into their theatres, and they say to one another, Why is it mourning? And they answer, These Jews observe the Sabbatical Year and, having no vegetables, have eaten its thorns, for which it mourns" (Lamentations Rabbah, Proems, xvii; and cf. *ibid.*, iii, 5). That the observance of the Sabbatical Year was still current among the Jews at that time is evident from this comment.

of the Torah was the tendency to establish an exclusive class of Sages enjoying special privileges by reason of their engaging in its study. This created a social stratification, whose basic principle was the study and knowledge of the Torah, with the talmid ḥakham, learned in the Torah, accorded the highest status, and the ʿam ha-aretz, ignorant in the Torah and lacking any desire to engage in its study, assigned to the lowest rank in society.

Against this background a hatred was engendered between the talmidei ḥakhamim and the ʿammei ha-aretz both because of the animosity which invariably prevails between the intellectual and the ignoramus, and because of the enmity which exists between the privileged upper and the despised and disparaged lower class.[60]

The period under discussion has bequeathed to Jewish society throughout all generations up to the present day this concept of the ʿam ha-aretz la-Torah, even as it has bequeathed the notion of study in general and of the study of the Torah in particular as the supreme value in Judaism. In all periods the expression ʿam ha-aretz has been used as a disparaging designation of the ignoramus who, unfamiliar with the sacred literature, the halakhot, and the civil laws, is contrasted with the erudite scholar of the Torah, who is learned in all this.

3. The Nature of the Social Stratum Known as the ʿAmmei ha-Aretz

An examination of the inquiries into the subject of the ʿam ha-aretz reveals various attempts at defining the concept socially. Some scholars saw in the ʿammei ha-aretz villagers as opposed to city-dwellers, others perceived in them farmers as against traders, some discerned in them a part of the early Christians as opposed to those who remained loyal to Judaism, others again viewed them as a sect alongside the sects of the Pharisees, the Sadducees, the Essenes, and so on.

But these distinctions are incompatible with the social situation that obtained in the days of the Second Temple and in the tannaitic period. Nor do they convey the social nature of the ʿam ha-aretz.

The distinction between cities and villages made in various studies on the present subject is a projection of a later situation upon the period of the Second Temple and of the Mishnah. Most settlements, some

---

[60] See below, pp. 170–199.

with no more than several hundred inhabitants, were called cities. Few had a population of many or tens of thousands. Only on this basis can we explain certain statements, such as, ועוד זאת היתה ירושלים יתרה על יבנה, שכל עיר שהיא רואה ושומעת וקרובה ויכולה לבא — תוקעין "Jerusalem had this further superiority over Jabneh: in every city, from which it could be seen or heard, and which was near, and from which it was accessible, they used to blow the shofar [on the Sabbath]."[61] The reference here is to "cities" within sight, within earshot, as well as within the Sabbath limits of Jerusalem, and with an unobstructed passage for anyone walking between them and Jerusalem. Obviously cities in the later, accepted sense of the word are not meant here, nor are they in the statement, תנו רבנן, פעם אחת חל ראש השנה להיות בשבת, והיו כל הערים מתכנסין [ליבנה] "Our Rabbis taught, Once the New Year fell on a Sabbath and all the cities gathered together [at Jabneh]."[62]

A settlement was apparently designated a city when it had its own public institutions and had no longer to rely on those of other settlements.[63]

Accordingly any distinction between rural farmers and urban craftsmen and traders is artificial. In the period of the Second Temple and of the Mishnah a city also had, or consisted only of, agricultural lands. Many city-dwellers owned fields. That a city was sometimes synonymous with a farm[64] is also borne out by a reference in the sources to עיר של יחיד "a city which belongs to an individual."[65] This identity

---

[61] Ro'sh ha-Shanah iv, 2.

[62] TB Ro'sh ha-Shanah 29b. In some MSS. however the word "cities" does not occur; see Diḳduḳei Soferim; cf. also ונכנסו כל העיירות להספידו "And all the cities assembled to mourn over him" [on the death of R. Judah ha-Nasi at Sepphoris] (TJ Ketubbot xii, 35a).

[63] On the nature of a city's public institutions, see TB Sanhedrin 17b; TJ Ḳiddushin iv, 66b; Seder Eliyahu Zuṭa, Pirḳei Derekh Eretz i, ed. Friedmann, p. 13; Tosefta Bava Metzi'a xi, 23; TB Bava Batra 8b; Seder Eliyahu Rabbah xi, ed. Friedmann, p. 54; etc. See S. Safrai, "Ha-'Ir ha-Yehudit be-Eretz Yisra'el bi-Teḳufat ha-Mishnah ve-ha-Talmud," *Ha-'Ir ve-ha-Ḳehillah*, Historical Society of Israel, Jerusalem, 1967, pp. 227–236; M. Weinberg, "Die Organisation der jüdischen Ortsgemeinden in der talmudischen Zeit," *MGWJ*, 41 (1897), pp. 588–604, 639–660, 673–691.

[64] For a city in the sense of a farm, see also TJ Yevamot viii, 8d: ר׳ יצחק בר נחמן בשם ר׳ יהושע בן לוי מעשה באחד שלקח עיר אחד של עבדים ערלים מן הגוי על מנת למוהלן וחזרו בהן "R. Isaac bar Naḥman in the name of R. Joshua b. Levi: A man once bought a city of uncircumcised slaves from a gentile on condition that he would circumcise them, but they retracted." And see TB Yevamot 48b; see also Bava Batra iv, 7: המוכר את העיר "If a man sells a city."

[65] 'Eruvin v, 6.

between a city and a farm occurs in the Bible too, as, for example, Havvoth-jair [lit. farms of Jair], which were also cities.[66]

There is thus clearly no reason to identify the 'ammei ha-aretz with villagers and/or with farmers and to see in them either villagers as opposed to city-dwellers, or farmers as opposed to merchants.

Scholars who identified the 'ammei ha-aretz with farmers did so because the 'am ha-aretz is defined as one who was not scrupulous in observing the commandments associated with the produce of the land in Eretz Israel. These commandments constituted a heavy burden, and there were apparently numerous reasons why the 'ammei ha-aretz did not observe them properly. The argument that the 'ammei ha-aretz did not separate tithes since this obligation was not imposed upon the merchants is not among the most plausible arguments to account for their evading the commandment. Moreover most of the halakhot dealing with the 'am ha-aretz le-mitzvot refer to him as suspected of not observing the laws of ritual purity, and this has nothing to do with villages and with agriculture.

Nor is there any justification for regarding the 'am ha-aretz as necessarily belonging to a lower or "plebeian" class. The majority of the 'ammei ha-aretz probably did belong to the masses, but there was no reason why some of them should not have been wealthy and members of the aristocratic and upper classes. The Mishnah refers to ממזר תלמיד חכם וכהן גדול עם הארץ "a bastard who is a talmid ḥakham and a high priest who is an 'am ha-aretz."[67] If, by definition, the 'am ha-aretz belonged to a low and inferior social class, "a high priest who is an 'am ha-aretz" is inconceivable, even as an absurd example.

The halakhot which deal with a ḥaver and an 'am ha-aretz or with a talmid ḥakham and an 'am ha-aretz reflect historical conditions in which the ḥaver or the talmid ḥakham did not invariably belong to a higher economic or "professional" class than the 'am ha-aretz. Thus the halakhah refers to חבר שהיה חוכר מעם הארץ "a ḥaver who hired from an 'am ha-aretz,"[68] הטוחן אצל עם הארץ "one who mills at an 'am ha-aretz,"[69] בן חבר שלמד אצל עם הארץ "the son of a ḥaver who studied under an 'am ha-aretz,"[70] and so on. We find 'ammei ha-aretz

66 Jos. 13:30; and cf. Jud. 10:4; I Kings 4:13; I Chron. 2:23.

67 Horayot iii, 8; and parallel passages.

68 Tosefta Demai iii, 5.

69 Tosefta Demai iv, 27.

70 Tosefta Demai ii, 18.

owning property, using shops together with ḥaverim,[71] living close to ḥaverim,[72] and selling male and female slaves to ḥaverim.[73] From these examples it can readily be seen that the 'am ha-aretz was not necessarily depicted as belonging to the masses and the ordinary people, in contrast to the upper classes that sought to humiliate him.[74]

Mention has previously been made of the attempt of scholars to see in the 'ammei ha-aretz a sect with its own principles, like the Essenes, the Christians, the ḥaverim, and others.

Individual halakhot seem indeed to indicate that the 'ammei ha-aretz constituted a sect. There is, for example, the advice given to the talmid ḥakham that ואל יסב בחבורה של עמי הארץ "he should not take a set meal in the company of the 'ammei ha-aretz,"[75] or the reference to ישיבת בתי כנסיות של עמי הארץ "sitting in the synagogues of the 'ammei ha-aretz."[76] That the 'ammei ha-aretz were a sect which had institutions of its own and which was conducted according to specific and established principles cannot be inferred from these statements, which should rather be interpreted as advice to a talmid ḥakham not to associate with 'ammei ha-aretz sitting together after the manner of those who sit in "the seat of the scornful." The reference is thus not to a closed association. Nor is the statement which mentions synagogues to be taken as evidence that the 'ammei ha-aretz had special synagogues of their own, conducted differently from others, but rather that the 'ammei ha-aretz by their behaviour, their non-observance of commandments, and/or their lack of Torah detracted as it were from the sanctity and uniqueness of the synagogues they frequented.

To sum up. The 'ammei ha-aretz are to be viewed as constituting a social stream which belonged to no defined class and had no separate organizational framework. Nor is the 'am ha-aretz to be seen as one

71 Tosefta Demai iii, 9.

72 Tosefta Ṭohorot ix, 1, 11; TB 'Avodah Zarah 70b.

73 Tosefta 'Avodah Zarah iii, 9.

74 On this, see below, pp. 161–169.

75 TB Berakhot 43b. Cf. also תלמידי חכמים נאים בחבורה, ואין עמי הארץ נאים בחבורה "Talmidei ḥakhamim are pleasant in company, and 'ammei ha-aretz are unpleasant in company" (Massekhet Derekh Eretz iii, 1, ed. Higger, p. 97).

76 Avot iii, 10, in the name of R. Dosa b. Harkinas. In the Kaufmann and Parma MSS., as well as in the Mishnayot from the Genizah, Antonin Collection (Ginzei Mishnah, p. 111), etc., the reading is ישיבת כנסיות של עמי הארץ "sitting in the assemblies of the 'ammei ha-aretz." And see below, pp. 170–199, and, especially, p. 173.

who intended to deny his Judaism or who basically refused to accept the principles of the Written or the Oral Law.

In the sources the ʻammei ha-aretz are defined as such because they did not scrupulously observe certain commandments and because of their ignorance of the Torah. These definitions and their social implications as far as the ʻammei ha-aretz are concerned, as also the relations of the ʻammei ha-aretz with the other sections of the nation, are discussed in the following pages.

CHAPTER TWO

TITHES AND RITUAL PURITY

A. TITHES

1. Introductory Remarks

The halakhah, as derived from talmudic sources, prescribed that a Jew who wished to make his produce fit for eating as secular food had to separate terumot [תרומות], or priestly dues, and tithes [מעשרות] in the following manner. First, he separated terumah for the priests. From the remainder he gave a tenth, which was the first tithe [מעשר ראשון], to the Levites, who for their part gave to the priests a tenth of the first tithe, called the terumah of the tithe [תרומת מעשר] or the tithe of the tithe [מעשר מן המעשר]. Having separated the terumah and the first tithe, the owner of the produce separated from the remainder yet another tenth, this being the second tithe [מעשר שני] in the first, second, fourth, and fifth years and the poor man's tithe [מעשר עני] in the third and sixth years of the Sabbatical cycle. The second tithe had to be taken to Jerusalem and consumed there, or redeemed and the money, with the addition of 25% [חומש מלבר], taken to Jerusalem where the owner bought whatever he wished. The poor man's tithe, given, as its name implies, to the poor, consisted of secular food. On the last day of Passover in the fourth and seventh years a confession, contained in the Bible (Deut. 26:13–15) and relating to the tithe, was made. The confession referred, according to the Sages, to all the tithes.

2. The Relation between the Biblical Law and the Halakhah

The halakhot mentioned in the previous section are based on the Bible, which contains several laws dealing with the tithes.

There are two such laws in the Priestly Code. Of these two, a detailed one occurs in the Book of Numbers, as follows. "To the Levites I have given every tithe in Israel for an inheritance, in return for their service which they serve, their service in the tent of meeting... And the Lord said to Moses, 'Moreover you shall say to the Levites, "When you take from the people of Israel the tithe which I have given you from them for your inheritance, then you shall present an offering from it to the Lord, a tithe of the tithe. And your offering shall be reckoned to you as though it were the grain of the threshing floor, and as the fulness of the wine press... and you may eat it in any place, you and your households; for it is your reward in return for your service in the tent of meeting...'' ' '' (Num. 18:21–32).

That the tithe was given to the Levites both in return for their service in the Tabernacle and because no inheritance of land would be allotted to them is evident from this law, which imposed on them the obligation to separate, from the first tithe given to them, the tithe of the tithe as the priestly terumah.

The halakhah sees in this law the basis of the first tithe which was given to the Levites.

About its date and significance biblical scholars disagree. Thus Wellhausen and others,[1] who assign source P to as late as the time of the Babylonian exile, hold that the law deals with the tithe given to the Levites as an annual obligation. They point to the identity between the regulations relating to the tithe as stated in the halakhah and those in the law under discussion, an identity which, they contend, is due to the proximity of time between the biblical law composed during the Babylonian exile and the halakhah which had its beginnings at the commencement of the Second Temple period.

Kaufmann,[2] seeking to prove the antiquity of the Priestly Code, maintains that the law of the tithe in the Book of Numbers refers to a tithe given either as a votive or as a freewill offering at no fixed date and with no time limit set for it. Since basically a votive or a freewill offering was dedicated to the Temple, the Levites had to separate the terumah of the tithe from the tithe in order to release the latter from its sanctity and so permit them to eat it as secular food anywhere without profaning ''the holy things of the people of Israel'' (Num. 18:32).

We shall not enter here into the question either of the date of the Priestly Code in particular or of the separation of terumot and tithes and the status of the priests and the Levites in the biblical period in general, all this being irrelevant to our present discussion. What is

---

1 J. Wellhausen, *Prolegomena zur Geschichte Israels*, Berlin, 1905[6], pp. 150–152.

2 Y. Kaufmann, *Toledot ha-Emunah ha-Yisra'elit*, Jerusalem & Tel Aviv, 1936–1956, I, pp. 143–184. The same view is expressed by M. Haran, *Entziklopedyah Mikra'it*, V, pp. 204–212, s.v. מעשר.

relevant is the relation between the biblical law and the halakhah. On this subject the views of the two schools are unacceptable.

There is no connection between the biblical law of the tithe and the halakhah of the Sages as far as the date of their formulation is concerned, since apparently the halakhah relating to the giving of the first tithe to the Levites is comparatively late. At the commencement of the Second Temple period there arose the tendency, which gained strength in the course of time, to give the first tithe to the priests [or also to the priests]. It is consequently difficult to comprehend how there could have been a contemporaneous biblical law prescribing that the tithe was to be given to the Levites only. Nor can it be reconciled with the sparse number of Levites in the early days of the Second Temple.[3] Nor does the view of Kaufmann and of others, who contend that the biblical law refers to a votive or freewill tithe which was not an annual obligation, fit in with the literal meaning of the biblical text. The very concept of the tithe is incompatible with a votive or a freewill offering. By its nature and meaning it refers to a fixed and defined obligatory tax, even as parallel cases of tithes among other peoples also refer to an annual obligatory tax.[4] Such is also the tax that is reflected in the verses, "To the Levites I have given every tithe in Israel," and "Moreover you shall say to the Levites, 'When you take from the people of Israel the tithe'."

The tithe mentioned in the Priestly Code also includes the innovation of the separation of the cattle tithe. "All the tithe of the land, whether of the seed of the land or of the fruit of the trees, is the Lord's; it is holy to the Lord. If a man wishes to redeem any of his tithe, he shall add a fifth to it. And all the tithe of herds and flocks, every tenth animal of all that pass under the herdsman's staff, shall be holy to the Lord. A man shall not inquire whether it is good or bad, neither shall he exchange it; and if he exchanges it, then both it and that for which it is exchanged shall be holy; it shall not be redeemed" (Lev. 27:30–33). This law declares that whereas the tithe of the seed and the fruit could be redeemed by the payment of the principal plus a fifth, the tithe of the herd and of the flock could not be redeemed. As against that mentioned in the Book of Numbers, the reference here is in all probability

[3] See p. 38, below.

[4] The ma'šharu in Akkadian texts from Ugarit, which was the tithe of the king, was an annual obligation, and in any event the king had the right to impose it as such. Among other peoples the same applied to a tithe for the king or the temple.

to a tithe given to the priest. This follows from the context, since the chapter as a whole deals with valuations, such as the firstling and the devoted thing, given to the priest for a sacrifice. The same apparently applies to the tithe of the cattle, given accordingly to the priests as a sacrifice of a peace-offering, the flesh of which was eaten by them and by the members of their households in ritual purity.

Even as biblical scholars hold different views on the law in the Book of Numbers, so do they disagree about the tithe of the cattle. Seeing in the tithe of the cattle a later addition and an extension of the law of the tithe in the Book of Numbers, Wellhausen and others argue that its late date is evident from the fact that there is no reference to it in the Book of Nehemiah in the passages dealing with the tithes, and that it is, therefore, later than Nehemiah. The tithe of the cattle is mentioned elsewhere only in the Book of Chronicles (II Chron. 31:6), where, it is true, the reference is to the days of Hezekiah, but the statement there apparently reflects the practice prevalent at the time of the Chronicler.

Kaufmann makes a distinction between the law of the tithe in Lev. 27 and that in the Book of Numbers. Thus he contends that, in contrast to the tithe in the Book of Numbers which was given to the Levite, all the tithe of the land and all the tithe of the animals were, according to the Book of Leviticus, holy, and hence belonged to the Temple or to the priest. Maintaining that this latter tithe is an ancient institution, Kaufmann finds its roots in the tithes of Abraham (Gen. 14:20) and of Jacob (Gen. 28:22), and compares with it also the tithe mentioned in the Book of Amos (4:4). Here, too, Kaufmann holds that the reference is not to an annual obligatory gift but to a voluntary dedication to the Temple in the form of a votive or a freewill offering.

In this instance we likewise have to disagree with these two schools.

The tithe of the cattle cannot be interpreted against the background of the halakhah current in the days of the Second Temple, since the halakhah did not recognize such a tithe given to the priest but rather understood the relevant verses as referring to a sacrifice which was to be designated as such at the peras[a] of the festival and the flesh of which was eaten by its owner in Jerusalem.[5] That this law is mentioned neither in Nehemiah nor in Malachi is no proof of its lateness, since the reference in these Books is to the tithe deposited in the Temple storehouses, and

---

[5] Zevaḥim v, 8; Bekhorot ix, 5; and cf. also TJ Sheḳalim iii, 47b; TB Bekhorot 58a.

a) Peras: lit. "half a month," that is, fifteen days before the festival.

while the tithe of the seeds and of the fruit could be stored there, the tithe of the cattle obviously could not. It is questionable whether the reference in the Book of Chronicles is indeed to the tithe of the cattle in a context in which we would above all expect the firstlings of the cattle to be mentioned but in which no reference is made to them. What may perhaps be meant there by the tithe of the cattle are the firstlings of the cattle, the word tithe being used because of the reference in the preceding verse to the tithe of the seeds and of the fruit. Several passages in the Apocrypha and in Philo[6] mention the tithe of the cattle but it is very doubtful whether these refer to a newly introduced practice. It seems rather that we have reflected here an earlier custom in its decline, or theoretical references to the law in Lev. 27, but not statements relating to prevailing conditions.[7] The very essence and nature of the tithe of the cattle suggest that it was introduced at an early stage in the history of the nation, seeing that it is a tithe appropriate to a pastoral society and hence compatible with a time when pasturage was a central component in the nation's economy.

Kaufmann's view that here, too, the reference is to a votive or a freewill tithe is also unacceptable for the same reasons that led us to reject his interpretation of the law of the tithe in the Book of Numbers. Furthermore the words כל אשר יעבור תחת השבט, "all that pass under the herdsman's staff," are consistent neither with a votive nor with a freewill offering but only with a compulsory tax, emphasizing as they do that there is no evasion of nor any substitution for it, that passing "under the herdsman's staff" determines which animal is holy to the Lord.

Whether Lev. 27 belongs to an early or to a late source, the intrinsic nature of the tithe of the cattle remains obscure. It may be conjectured that it refers to a law associated with a local sanctuary of a type that existed during most of the period of the settlement in Eretz Israel and

---

[6] Book of Jubilees 13:25–27; 32:15; Tobit [Codex Sinaiticus] 1:6–7; Philo, *De Virtutibus*, §95; and see also *idem*, *De Specialibus Legibus*, I, §141. Didache also mentions the tithe of the cattle that was given to the priests (13:3). The Karaites and Samaritans likewise laid down that the tithe of the cattle was to be given to the priest (*Sefer ha-Mivḥar* to Lev. 27:32; to Deut. 12:6; Anan, *Sefer ha-Mitzvot*, ed. Schechter, pp. 3 ff.; Samaritan Book of Joshua xxxviii in Kirchheim, *Karmei Shomeron*, pp. 75–76; and see Ch. Albeck, *Das Buch der Jubiläen und die Halacha*, Berlin, 1930, pp. 30, 57).

[7] Most of the apocryphal sources mentioned reflect conditions at the beginning of the Second Temple period. This also applies to Philo, who often cites a tradition which mirrors a custom prevalent before his days.

of the kingdom in biblical times. If this is so, we can understand the fragmentary, obscure mention of the law in Lev. 27, a law unknown to the other texts that deal with the subject of the tithe.

Entirely different from the law of the tithe in the Priestly Code (P) is that in the Book of Deuteronomy (D).

"You shall tithe all the yield of your seed, which comes forth from the field year by year. And before the Lord your God, in the place which he will choose, to make his name dwell there, you shall eat the tithe of your grain, of your wine, and of your oil, and the firstlings of your herd and flock... And if the way is too long for you, so that you are not able to bring the tithe... because the place is too far from you, which the Lord your God chooses to set his name there, then you shall turn it into money, and bind up the money in your hand, and go to the place which the Lord your God chooses, and spend the money for whatever you desire, oxen, or sheep, or wine or strong drink, whatever your appetite craves; and you shall eat there before the Lord your God... And you shall not forsake the Levite who is within your towns, for he has no portion or inheritance with you. At the end of every three years you shall bring forth all the tithe of your produce in the same year, and lay it up within your towns; and the Levite, because he has no portion or inheritance with you, and the sojourner, the fatherless, and the widow, who are within your towns, shall come and eat and be filled..." (Deut. 14:22–29).

The law of the tithe in the Book of Deuteronomy refers explicitly to an annual obligatory tithe to be separated from the increase of the seed and from the fruit. This tithe, or the food bought with its redemption money, was to be eaten by the owner at the Sanctuary. Every third year, when it was not obligatory to bring the tithe to Jerusalem, it could be eaten anywhere. In that year, the prayer stated in Deut. 26:12–15 was recited in Jerusalem. According to its literal meaning, the purpose of the prayer was to release the tithe from its holiness and permit it to be eaten anywhere, even as the Levite's separation of the tithe of the tithe released his tithe from its sanctity, enabling him to eat it wherever he wished. It is stressed that each year some of the tithe was to be given to the Levites [apparently to those who were not then serving in the Sanctuary — "the Levite who is within your towns"], and in the third year also to the sojourner, the fatherless, and the widow.

The halakhah does not connect the tithe of the Book of Deuteronomy with that of Num. 18 [which, according to the halakhah, deals with the first tithe], for according to the halakhah the Book of Deuteronomy refers to the second tithe which was separated from the produce after the separation of the first tithe and which was eaten in Jerusalem by its owner. The redemption of the second tithe is dealt with in Lev. 27: 30–31, the halakhah prescribing that if it was redeemed, 25% [חומש מלבר] of its value had to be added to it.

The tithe of the third year mentioned in the Book of Deuteronomy is explained by the halakhah as referring to the poor man's tithe, to be separated in the third and sixth years of the Sabbatical cycle.[8]

The prayer relating to the "removal" of the tithes, which occurs in Deut. 26, was called by the Sages וידוי מעשר, the confession or solemn declaration that not only the tithe of the third year had been duly given and removed from the householder's possession but also all tithes, as well as the terumah, the fruit of trees in their fourth year [נטע רבעי], and the priest's portion of the dough [חלה].[9]

To sum up the relation between the halakhah and the biblical law: the halakhah dealt with the laws of the tithe in the Priestly Code and in the Book of Deuteronomy, which it saw as referring to tithes that were different but yet had in common the fact that they constituted an annual obligation.

3. The Relation between the Theoretical Halakhah and the Prevailing Situation

Between the theoretical halakhah, as set forth in talmudic literature, and the prevailing situation relating to the separation of the tithes in the days of the Second Temple and of the Mishnah there is a gap, a circumstance which has a basic connection, as we shall show in the course of our remarks, with the 'am ha-aretz and with the problem of the disregard, and the lack of a scrupulous observance, of the separation of the tithes. With this gap we shall deal in its relation to the following subjects.

8 In several non-talmudic halakhic sources the order of the tithes is given differently: the first and second tithes are separated every year, while in the third and sixth years of the Sabbatical cycle there is, besides these, also the tithe of the poor (Book of Jubilees 32:11; Josephus, *AJ*, iv, 8, 22, §240). On the other hand, Tobit [Codex S.] 1:7–8, as well as the LXX to Deut. 26:12, correspond to the halakhah.

9 Ma'aser Sheni v, 10; Sifrei Deut., Ki Tavo, 26:12 ff., cccii ff., ed. Finkelstein, pp. 320 ff.; etc.; etc. According to the halakhah, the confession was made on the last day of Passover in the fourth and seventh years.

a) How the tithes were separated and where they were given.
b) The right of the priests and of the Levites to the tithes.
c) The tithes after the destruction of the Second Temple.
d) The separation of tithes in the diaspora.

a) *How the Tithes Were Separated and Where They Were Given*

According to the theoretical halakhah a man had the right to give the first tithe to a Levite[10] — to any Levite wheresoever he wished[11] [in a similar manner the halakhah defines a person's rights with regard to any of the other gifts of the priests and of the Levites].

Under the conditions prevailing during most of the Second Temple period this halakhah was not observed. Several scholars[12] have pointed out that numerous sources indicate and clearly prove that in the days of the Second Temple the custom had taken firm root of bringing the first tithe [as well as the other priestly and levitical gifts] to Jerusalem and the Temple, where it was distributed among the priests and the Levites.

In the days of the First Temple, too, it was apparently obligatory to bring the tithes to Jerusalem. This is attested to primarily by the law of the tithe in the Book of Deuteronomy [although according to the halakhah the reference there is to the second tithe]. In this manner we should also understand verses such as, "And thither you shall bring your burnt offerings and your sacrifices, [and] your tithes" (Deut. 12:6; see also verse 11), or, "Come to Bethel, and transgress; to Gilgal, and multiply transgression; bring your sacrifices every morning, your tithes every three days" (Amos 4:4; this is also clearly laid down, in connection with the first-born, in Ex. 34:20).

The biblical books dating from the Second Temple period speak explicitly of the bringing of the tithes, together with the other priestly gifts, to Jerusalem and their distribution there. The Books of Nehemiah,

---

10 See below, pp. 38–42, where it is pointed out that the right to the first tithe was granted in the Second Temple period also, or perhaps mainly, to the priests.

11 Sifrei Deut., Re'eh, lxxvii, ed. Finkelstein, p. 143; Midrash Tannaim to Deut. 14:22, p. 77; Tosefta Pe'ah iv, 7.

12 A. Büchler, "He'arot ve-He'arot 'al Matzav ha-Ishah be-Sefer Yehudit," *Sefer Blau*, Budapest, 1926, pp. 53 ff.; P. Churgin, "Sefer Yehudit," *Meḥḳarim bi-Teḳufat Bayit Sheni*, New York, 1949, pp. 130–132; G. Allon, *Meḥḳarim be-Toledot Yisra'el*, Tel Aviv, 1957–1958, I, pp. 83 ff.; S. Safrai, *Ha-'Aliyah le-Regel bi-Yemei Bayit Sheni*, Tel Aviv, 1965, pp. 127 ff.; S. Belkin, *Philo and the Oral Law*, Cambridge (Mass.), 1940, pp. 70 ff.

Malachi, and Chronicles record instances of the bringing of the tithes to the chambers [לשכות] and storehouses [אוצרות, בית אוצר] of the Temple.[13]

Accounts of the bringing of the tithes to Jerusalem are contained in non-talmudic halakhic sources. Thus the Book of Judith mentions the keeping of the tithes with the object of taking them later to Jerusalem,[14] while Tobit, who eulogizes his own righteousness, tells that he brought the tithes to Jerusalem.[15] What is more, from the Book of Maccabees we learn that when it was impossible to bring the tithe to Jerusalem, the commandment of the tithe could not be fulfilled. καὶ ἤνεγκαν τὰ ἱμάτια τῆς ἱερωσύνης καὶ τὰ πρωτογενήματα καὶ τὰς δεκάτας· καὶ ἤγειραν τοὺς ναζιραίους, οἳ ἐπλήρωσαν τὰς ἡμέρας· καὶ ἐβόησαν φωνῇ εἰς τὸν οὐρανὸν λέγοντες τί ποιήσωμεν τούτοις, καὶ ποῦ αὐτοὺς ἀπαγάγωμεν. "They brought also the priests' garments, and the first fruits, and the tithes; and the Nazirites they stirred up, who had accomplished their days. Then cried they with a loud voice toward heaven, saying, What shall we do with these, and whither shall we carry them away?" (I Macc. 3:49–50).

Even as it was obligatory to bring to the Temple the first fruits, the priests' garments,[16] and the sacrifices of the Nazirite, so the taking of the tithes to the Temple and their distribution there constituted the only way of observing the commandment of the tithe. To I Sam. 1:21 ["And the man Elkanah and all his house went up to offer to the Lord the yearly sacrifice, and to pay his vow"] the LXX added καὶ πάσας τὰς δεκάτας τῆς γῆς αὐτοῦ: "and all the tithes of his land." In no MS. or version of the Book of Samuel do these words occur, and they are therefore, so it seems, a later addition which reflects the practice current in the days of the translator.[17] Philo too speaks of bringing

---

[13] Neh. 13:5, 12–13; Mal. 3:10; II Chron. 31:5–12: although this passage refers to the days of Hezekiah, it presumably reflects the practice prevalent at the time the Book was composed.

[14] Judith 11:13.

[15] Tobit 1:6–7.

[16] On the bringing of the priests' garments to Jerusalem mentioned in this source in the Book of Maccabees, cf., for example, Tosefta Yoma i, 23; and see A. Büchler, *Die Priester und der Cultus im letzten Jahrzehnt des jerusalemischen Tempels*, Vienna, 1895, p. 200; Safrai, *Ha-'Aliyah le-Regel*, p. 50.

[17] A similar addition occurs in Josephus, *AJ*, v, 10, 3, §346, where it is stated that after the birth of Samuel, Elkanah and Hannah came to Shiloh to offer a sacrifice and to bring their tithes. In this, Josephus undoubtedly followed the LXX [the Vulgate also has this addition].

the tithes and the other priestly gifts to Jerusalem and to the Temple,[18] and explains the reason for their being brought to Jerusalem as intended to make the priests realize that they received their gifts not from man but from Heaven, and accordingly had no need to feel any shame. Too much significance should not, however, be attached to this explanation, which is but another of those instances in which Philo gave to an existing halakhah his own philosophical interpretation, influenced as a rule by Stoic doctrines.[19]

From these sources it is evident that at the beginning of the Second Temple period[20] it was usual to bring the tithes to the Temple and to distribute them there among those who had a right to them. Instituted apparently by Ezra and Nehemiah, this procedure is to be seen as part of their general tendency to advance the status of Jerusalem. The bringing of the tithes there reinforced the city both nationally and economically, and enabled the priests and the Levites to serve in the Temple. That the tribe of Levi had no inheritance is given in the biblical law as the reason for the commandment of the tithes. Yet in the days of the Second Temple the priests and the Levites had in fact landed property just like the rest of the people. It is accordingly understandable why there should have been a tendency to bring the priestly and levitical gifts to the Temple and to distribute them there proportionately according to the needs both of the Temple and of the priests and Levites then officiating in their division [משמר].[21]

No specific time was presumably laid down for taking the tithes to

---

[18] Philo, *De Specialibus Legibus*, I, §§132–152.

[19] Scholars also cite the statement of Demetrius in his letter "unto the nation of the Jews" as proof that the tithes were brought to Jerusalem: καὶ Ιερουσαλημ ἤστω ἁγία καὶ ἀφειμένη καὶ τὰ ὅρια αὐτῆς, αἱ δεκάται καὶ τὰ τέλη "Let Jerusalem also be holy and free, with the borders thereof, both from tithes and tributes" (I Macc. 10:31; cf. also Josephus, *AJ*, xiii, 2, 3, §51). These scholars base themselves on the translations and comments *ad loc.* of Kautzsch and Charles. It is however more probable that the term "tithes" in this passage does not refer to the first tithe but to the government taxes, from which Demetrius exempted the inhabitants of Jerusalem. If this is so, the verse cannot be used as proof in the subject under discussion. And see M. Stern, *Ha-Te'udot le-Mered ha-Ḥashmona'im*, Tel Aviv, 1965, p. 102.

[20] On the assumption that the Book of Judith was composed in the Persian period (Y. M. Grintz, *Sefer Yehudit*, 1957, pp. 3–17; etc.) and that the Book of Tobit is not later than the third century BCE (see D. Flusser, *Entziḳlopedyah Miḳra'it*, III, pp. 367–375, s.v. טוביה, and the bibliography *ad loc.*). Philo also sometimes reflects the ancient halakhah.

[21] See Allon, *Meḥḳarim*, I, p. 87.

Jerusalem. Naturally they were brought there during a pilgrimage.[22] Some of the sources that attest to the bringing of the tithes to Jerusalem do so either directly or indirectly.[23] The Mishnah Ḥagigah, which states that ובשעת הרגל [נאמנים] אף על התרומה "during a festival [they are trusted] also in respect of the terumah,"[24] likewise reflects a situation in which the terumah was brought to Jerusalem on the occasion of a festival, terumah being specifically mentioned, and not tithes, undoubtedly because of its holiness. Also brought to Jerusalem at the time of a festival was the tithe of the cattle.[25]

The Jerusalem Talmud cites a tradition in the name of R. Joshua b. Levi which apparently describes the manner in which the tithe was distributed in former times. בראשונה היה מעשר נעשה לשלשה חלקים, שליש למכרי כהונה ולוייה, ושליש לאוצר, ושליש לעניים ולחבירים שהיו בירושלים "At first the tithe was divided into three parts, one-third was for the friends of the priests and the Levites, one-third for the treasury, and one-third for the poor and the ḥaverim in Jerusalem."[26] This statement of the Jerusalem Talmud occurs in a context in which the ordinances of Johanan the high priest [John Hyrcanus][27] are explained, and the expression "at first" denotes that, in the view of the author, the statement reflects a situation and a regulation that obtained prior to the ordinances of Johanan the high priest.

The actual distribution of the tithe mentioned in the statement is

---

22 See *ibid.*, p. 85; and, in detail, Safrai, *Ha-'Aliyah le-Regel.*

23 In the Book of Judith mention is made of keeping the tithes in order to bring them to Jerusalem. Hence they were not brought to Jerusalem soon after being separated but almost certainly when the pilgrimage was made to Jerusalem. This is clearly stated in Tobit: "But I alone used often to journey to Jerusalem at the feasts... with the first fruits... and the tenth of the corn and the wine and oil..." (Tobit 1:6–7). In the Book of Samuel, too, in the passage in which the LXX added a reference to the bringing of the tithes to the Sanctuary, this took place apparently at the time of a pilgrimage when Elkanah went to Shiloh "to offer to the Lord the yearly sacrifice" (I Sam. 1:21).

24 Ḥagigah iii, 6.

25 This follows from the Book of Tobit [Codex S.] and this, too, is the view of R. Akiva in Bekhorot ix, 5: שלש גרנות למעשר בהמה, בפרס הפסח, בפרס העצרת, בפרס החג – דברי ר' עקיבא "There are three periods for the tithe of cattle. In the peras [lit. 'half a month,' i.e. fifteen days before the festival] of Passover, in the peras of Pentecost, in the peras of Tabernacles. These are the words of R. Akiva." The reason is given in the Talmud: כדי שתהא בהמה מצויה לעולי רגלים "So that cattle may be available for pilgrims" (TJ Sheḳalim iii, 47b).

26 TJ Ma'aser Sheni v, 56d; TJ Soṭah ix, 24a.

27 Ma'aser Sheni v, 15; Soṭah ix, 10.

somewhat obscure. Most probably "the friends of the priests and the Levites" refer to the priests and the Levites of the division [משמר] then serving in Jerusalem,[28] "the treasury" refers to the Temple treasury or, if the statement dates from the Hasmonaean period, the state treasury, and "the poor and the ḥaverim" refer to the priestly ḥaverim or the poor who were ḥaverim, the tithes being given to them because of their need and because, being ḥaverim, they were scrupulous in eating them in ritual purity.

The Hasmonaean kings sought to take control of the tithes and to use them for their own purposes.[29] It is against this background that we can understand the ordinances of Johanan the high priest. יוחנן כהן גדול העביר הודית המעשר. אף הוא בטל את המעוררים, ואת הנוקפים. ועד ימיו היה פטיש מכה בירושלם. ובימיו אין אדם צריך לשאול על הדמאי "Johanan the high priest set aside the confession of the tithes. He also abolished the 'wakers' and the 'strikers.' Until his days the hammer used to beat in Jerusalem. And in his days one had no need to inquire concerning demai." These ordinances were apparently related to the arrangements for collecting the tithes and concentrating them in the hands of the government. Because of these arrangements there was no longer any need "to inquire" concerning demai seeing that the tithes were taken from everyone. Nor was there any purpose in continuing the confession relating to the tithes once they were levied as a tax by the Hasmonaean authorities. It is accordingly understandable how it came about in the days of John Hyrcanus that, on the one hand, the confession that all the tithes had been given was abolished, while, on the other, no misgivings were entertained about demai, seeing that the separation of the tithe was now imposed by the government. Of this situation an echo can be found in the tradition in the Jerusalem Talmud: ובימיו אין אדם צריך לשאול על דמאי – שהעמיד זוגות "And in his days one had no need to inquire about demai, for he instituted pairs,"[30]

---

[28] Cf. the expression יהויריב ומכיריו "Jehoiarab and his friends": TJ Bikkurim iii, 65c. See S. Lieberman, "Emendations on the Jerushalmi" (Hebrew), *Tarbiz*, 3 (1932), pp. 210–212; unlike Allon, *Meḥḳarim*, I, p. 89, note 18, and Safrai, *Ha-'Aliyah le-Regel*, pp. 128–129. According to them, this third was given to the priests and Levites in the place where the owner of the produce resided. Technically it is difficult to assume that the owner split up the giving of the tithes in this way, handing out one-third where he lived and taking two-thirds to Jerusalem.

[29] A. Schalit, *König Herodes, der Mann und sein Werk*, Berlin, 1969, pp. 262–271, and note 424.

[30] TJ Ma'aser Sheni v, 56d.

these being state-appointed tithe collectors who worked in pairs.

The other ordinances of Johanan the high priest mentioned in the Mishnah may also be related to the subject of the levying of the tithes. "The wakers, the strikers, and the hammer that used to beat" were Ptolemaic methods of levying taxes. When John Hyrcanus appointed collectors to levy the tithes directly in accordance with a state decree, the remnants of the Ptolemaic method, being less efficient than the system introduced by John Hyrcanus, could be abolished.[31]

That the tithes were brought to Jerusalem and that the Hasmonaean rulers used them for their own purposes can be clearly seen from Julius Caesar's edict. Issued when Julius Caesar came to Eretz Israel and instituted a new order to replace the arrangements of Gabinius, this edict several times mentions the subject of the tithes. By it, Julius Caesar conferred on Hyrcanus the status of an ethnarch, granted him additional privileges,[32] and further declared: πρὸς τούτοις ἔτι καὶ Ὑρκανῷ καὶ τοῖς τέκνοις αὐτοῦ τὰς δεκάτας τελῶσιν, ἅς ἐτέλουν καὶ τοῖς προγόνοις αὐτῶν. "And in addition, they shall also pay tithes to Hyrcanus and his sons, just as they paid to their forefathers" (*AJ*, xiv, 10, 6, §203).[33] Caesar emphasized that these were not new privileges, since Hyrcanus' right to the tithes derived from the rights enjoyed by his forefathers. It is thus clear that under the Hasmonaeans the tithes were brought to Jerusalem and that the right to them and to distribute them belonged to the Hasmonaean ruler.

Thus in the Persian and in the Hellenistic periods, as also in the days of the Hasmonaeans, the custom was maintained of bringing the tithes to Jerusalem, and only the manner of utilizing them varied, as can be seen from the way in which they were levied and distributed. From the

---

31 Allon, *Meḥḳarim*, I, pp. 88–92, ascribes these enactments to Hyrcanus [II] and sees their background in Caesar's edict granted to Hyrcanus [see later]. Allon's view has been refuted. In general, it is not surprising that a similar subject [enactments arising from the failure to separate tithes properly] should be mentioned, and that they should have occurred at two different periods. But this is no reason for not ascribing the enactments to John Hyrcanus [I], as can also be seen from the Jerusalem Talmud (TJ Ma'aser Sheni v, 56d; TJ Soṭah ix, 24a). For details on the disregard of the obligation to separate tithes, see below, pp. 69–78.

32 Josephus deals at length with the grant of the privileges, presumably because of his persistence in defending the Jews before the Greeks and the Romans against the slander that they were forgers. Nonetheless it may be assumed tnat the kernel of the evidences is authentic.

33 For an explanation of the edict as a whole, see Schalit, *König Herodes*, Appendix XIII, pp. 777–781.

days of Ezra and Nehemiah onwards the tithes were brought to the Temple storehouses and there distributed among the priests and the Levites serving in the Temple. A later tradition, cited above and apparently referring to the early days of the Hasmonaeans, tells of the distribution of the tithes among the division of the priests and the Levites, the treasury, and the poor and the ḥaverim in Jerusalem. A decisive change is reflected in the ordinances of John Hyrcanus, who undoubtedly introduced an organized system of collecting the tithes, which were placed in the storehouses of the ruler and of the state, and if they were also given to the priests, presumably the relatives and close friends of the Hasmonaean ruler received them. At the beginning of the Roman period the edict of Julius Caesar officially reinforced and reconfirmed the custom whereby the tithes were brought to Jerusalem and distributed there by the ruler.

The manner in which the tithes were utilized by the Hasmonaean kings was unacceptable to the Pharisees who saw in it a clear indication of the people's enslavement. The concentration of the tithe money in the hands of the Hasmonaean kings contravened the biblical prohibition, "Nor shall he greatly multiply for himself silver and gold" (Deut. 17:17). Moreover this money was undoubtedly often used by the kings for purposes which did not always meet with the approval of the Pharisees. Against this background and that of the general desire of the Pharisees to harm the Hasmonaean kings there was presumably introduced the halakhah which we quoted at the beginning of this section and which states that the tithes were given directly to any Levite [or priest] wherever the owner of the produce wished. By this halakhah the Pharisees sought, on the one hand, to ensure the observance of the commandment of the separation of the tithes and, on the other, to withhold them from the Hasmonaean kings. Two evidences reinforce this assumption.

i) The entire Pharisaic halakhah makes no mention of bringing the tithes to Jerusalem,[34] although this is attested to by all the evidences from the days of the Restoration to the beginning of Roman rule. It can only mean that we have here a tendentious approach explicable on the basis of the Pharisees' opposition to the bringing of the tithes to Jerusalem where they would be concentrated in the hands of the Hasmonaean kings.

---

[34] The statement of the Amora R. Joshua b. Levi, quoted above, is the solitary, albeit obscure, evidence to the contrary.

ii) When Julius Caesar granted John Hyrcanus the right to the tithes, he in point of fact confirmed an ancient Jewish law. There was presumably no need for such a confirmation by a Roman ruler. Yet in view of the developments in and before the days of Hyrcanus the separation of the tithes was not strictly observed. This was either because they were not brought to Jerusalem or were not handed over to the state collectors but given instead to the Levites [or the priests] wherever the owners of the produce lived, as laid down by the new Pharisaic halakhah.[35]

Several evidences from the end of the Second Temple period reflect a situation, such as that permitted by the halakhah, whereby the tithes could be separated anywhere and given directly to the Levites [or priests]. Attesting to this are expressions such as כהנים ולויים שהיו עומדים על הגורן, "priests and Levites who stood on the threshing floor";[36] לאפוקי מעשריא, "to bring out the tithes";[37] and so on.

Josephus tells that his two fellow-priests, Joazar and Judas, who were sent with him to Galilee at the beginning of the Great Revolt, amassed a large sum of money from the tithes and wanted to return home.[38] Elsewhere, when praising his own conduct, he says that despite his rights as a priest, he declined to accept the tithes due to him from those who brought them to him.[39] In *Antiquities*, when telling of Agrippa II and of the high priest Ishmael the son of Phabi appointed by him, Josephus says that the high priests sent their slaves to the threshing floors to take the tithes by force, so that the ordinary priests, deprived of their due share, starved to death.[40] He gives a similar account of the

---

[35] Further proof may perhaps be found in the later statement that the separation of the tithes in the days of the Second Temple was not prescribed by biblical law, but that מאיליהן קיבלו "they voluntarily assumed the obligation" (TJ Shevi'it vi, 36b; TJ Ma'aserot ii, 49c; Gen. Rabbah xcvi, ed. Theodor-Albeck, p. 1202; Ruth Rabbah i, 4). This statement may be the consequence of the tradition about the Sages' innovations concerning the separation of tithes in the days of the Second Temple, or may even reflect the opposition to the Hasmonaeans' methods of levying the tithes when, in order to reinforce the Sages' authority to modify the laws of the tithes, it was laid down that these laws were not prescribed by the Bible.

[36] Tosefta Pe'ah iv, 3–6; etc.

[37] In Rabban Gamaliel's letter: Tosefta Sanhedrin ii, 6; and in the parallel passages: TJ Sanhedrin i, 18d; TB Sanhedrin 11b; Midrash Tannaim to Deut. 26:13, p. 176. This expression does not however constitute definite proof, since it may also refer to "taking out" the tithes so that the tithe collectors could take and bring them to Jerusalem.

[38] *Life*, xii, §§62–63.

[39] *Ibid.*, xv, §80.

[40] *AJ*, xx, 8, 8, §181.

high priest Ananias, who earned a high reputation for himself but whose evil servants were among those who forcibly took from the threshing floors the tithes of the priests.[41]

Since most of these evidences date from the last days of the Second Temple, it is difficult to determine whether the separation of the tithes anywhere was the result of fulfilling the Pharisaic halakhah or was due to the anarchy then reigning, for the farmer was clearly not free to give the tithes to whomsoever he wished but was under pressure from distinguished and violent priests. It seems therefore that it was only the destruction of the Second Temple which tipped the scale in favour of the Pharisaic halakhah permitting the owner of the produce to give the tithes wherever and to whatever Levite [or priest] he wished.

b) *The Right of the Priests and of the Levites to the Tithes*

According to the theoretical halakhah the first tithe was assigned to the Levites in accordance with the law of the tithe in the Book of Numbers. On the other hand there are numerous evidences that from the days of Ezra and Nehemiah the tithe was allotted also to the priests, and perhaps only to them.

The first time we come across tithes given to the priests in the Second Temple period is in the Book of Nehemiah. Although the "firm covenant" refers to the giving of the tithes to the Levites (Neh. 10:38–40), elsewhere in the Book the priests are also mentioned as receiving them. One passage even tells that the Levites fled from ministering in the Temple to their fields because they were not given their portion of the tithes (Neh. 13:4–13; see also 12:44).

The talmudic tradition, in referring to the right to the tithes granted to the priests in the days of Ezra and Nehemiah, explains this as due to the small number of Levites who returned to Zion.[42] And indeed the figures of those who came back as given in the Books of Ezra and Nehemiah confirm the talmudic statement about the small number of Levites as compared to priests.[43]

---

[41] *Ibid.*, xx, 9, 2, §§206–207.

[42] TB Yevamot 86b; cf. also TB Ketubbot 26a; TB Bava Batra 81b; TB Ḥullin 131b.

[43] The Levites who returned to Eretz Israel under Zerubbabel were about a tenth of the number of the priests: Ezra 2; Neh. 7:6. In the account of the decree that one-tenth of the population of the country were to live in Jerusalem, the number of the Levites was about a third of that of the priests: Neh. 11. Cf. also Neh. 3:17; chap. 6; and also the verse (Ezra 8:15) on which the Talmud bases itself.

In the apocryphal literature, in the passages referring to tithes, there are evidences that this situation, whereby the priests received tithes, continued throughout the Second Temple period. While the Books of Judith and of Jubilees explicitly refer to the tithes being given to the priests,[44] in the Book of Tobit the reference is obscure, mentioning as it does that the tithes were given to both the priests and the Levites: τὰς ἀπαρχὰς καὶ τὰς δεκάτας τῶν γενημάτων καὶ τὰς πρωτοκουρίας ἔχων· καὶ ἐδίδουν αὐτὰς τοῖς ἱερεῦσιν τοῖς υἱοῖς Ααρων πρὸς τὸ θυσιαστήριον πάντων τῶν γενημάτων· τὴν δεκάτην ἐδίδουν τοῖς υἱοῖς Λευι τοῖς θεραπεύουσιν ἐν Ιερουσαλημ, κτλ. "...with the first fruits [or terumot] and the firstlings and the tenths of the increase and the first shearings of the sheep, and [used to] give them to the priests, the sons of Aaron, for the altar, and the tenth of the corn and the wine and oil and pomegranates and the rest of the fruits to the sons of Levi, who ministered at Jerusalem..."[45] The duplication in these verses, which according to their literal meaning speak of the tithes first as included among the gifts of the priesthood and then as given to the Levites, may reflect the contradiction between the situation that existed in the days of Tobit, when tithes were given to the priests, and the law in the Book of Numbers which lays down that the tithes were to be given to the Levites. This duplication is due to Tobit's tendency to show how scrupulous he was in observing the commandments of the Torah.

That the tithes were given to the priests is also borne out by the LXX addition to I Sam. 1:21.

In some passages Philo and Josephus state that the tithes were given to the priests,[46] in some that they were given to both the priests and the

---

[44] καὶ τὰς ἀπαρχὰς τοῦ σίτου καὶ τὰς δεκάτας τοῦ οἴνου καὶ τοῦ ἐλαίου, ἃ διεφύλαξαν ἁγιάσαντες τοῖς ἱερεῦσιν τοῖς παρεστηκόσιν ἐν Ιερουσαλημ. "...The first fruits of the corn [or the terumot] and the tenths of the wine and the oil, which they had sanctified, and reserved for the priests that stand before the face of our God in Jerusalem" (Judith 11:13). "...That they should give to the Lord the tenth of everything, of the seed and of the wine and of the oil and of the cattle and of the sheep. And he gave [it] unto his priests to eat and to drink with joy before him" (Book of Jubilees 13:26–27).

[45] Tobit 1:6–7.

[46] Philo, *De Virtutibus*, §95; Josephus, *AJ*, xx, 8, 8, §181; xx, 9, 2, §§206–207; *idem*, *Life*, xii, §§62–63; xv, §80.

Levites,[47] in others again to the Levites only.[48] In this instance, too, the contradictions are to be explained as a confusion between the position that prevailed in the Second Temple period on the one hand and the biblical law in the Book of Numbers and perhaps also the halakhah based on that biblical law on the other.

Most scholars, relying on the above evidences, went so far as to assert that during the whole or part of the Second Temple period the tithes were given only to the priests. There is no definite proof of this, and it must be assumed that priests and Levites alike received a share of the tithes. Wherever the word "priests" is used in the sources quoted, as it is in most of them, the expression is not precise and could also include the Levites. So, for example, when the Book of Judith tells of the giving of the tithes to "the priests that stand before the face of our God in Jerusalem," the reference is presumably also to the Levites, since the expression "stand before the face" means ministering,[49] and the phrase "stand before the face of God" applied to both the priests and Levites who ministered in the Temple.[50] Nor from the practical aspect can it be assumed that the Levites, who did all the hard work in the Temple, would have continued to do it without some compensation. Also attesting to the distribution of the tithes among the priests and the Levites is the tradition in the Jerusalem Talmud, quoted above, that "one-third was for the friends of the priests and the Levites."

Tithes were also given to the priests in the days of the kings of the Hasmonaean dynasty who, as previously mentioned, took control of the tithes for their own needs. An echo of this may perhaps be found in an explanation given by the Talmud for the ordinance of Johanan the high priest. יוחנן כהן גדול העביר הודיית המעשר כו׳: מאי טעמא, אמר רבי יוסי בר׳ חנינא, לפי שאין נותנין אותו כתיקונו, דרחמנא אמר דיהבי ללוים, ואנן קא יהבינן לכהנים "Johanan the high priest abolished the confession that the tithe had been given. What was his reason? — R. Jose b. Ḥanina

[47] *Idem*, *AJ*, iv, 4, 3, §68; iv, 8, 8, §205. Albeck, in the introduction to tractate Ma'aserot in his commentary on the Mishnah, maintains that wherever Josephus says that the tithe was given to the priests and the Levites, he means the Levites, who then gave a tithe of the tithe to the priests. This view is however improbable, and was advanced in an attempt to reconcile Josephus' statement with the halakhah.

[48] Philo, *De Specialibus Legibus*, I, §156; Josephus, *AJ*, iv, 8, 22, §§240–243; iv, 4, 4, §69.

[49] Cf. II Kings 3:14; I Kings 10:8; II Chron. 9:7.

[50] Judith 4:14. Cf. also Zech. 3:7; Ps. 135:2; and, in particular, Deut. 18:7 — "Then he may minister in the name of the Lord his God, like all his fellow Levites who stand to minister there before the Lord."

said, Because people were not presenting it as prescribed. For the Divine Law states that it is to be given to the Levites, whereas we give it to the priests."[51] Since the tithe was presented to the priests, the confession that it had been duly given could no longer be made, seeing that it included the words, "And moreover I have given it to the Levite" (Deut. 26:13). Johanan the high priest therefore abolished the confession.[52]

The Pharisaic halakhah which states that the first tithe was given only to the Levite may have crystallized in a manner similar to the halakhah which, in opposition to the Hasmonaean kings and to their taking control of the tithes, permitted the tithes to be given anywhere and not necessarily brought to Jerusalem. On this assumption we can understand why in this instance, too, there is almost no reference in the theoretical halakhah to the giving of the tithes to the priests, even though this was done during the entire period of the Second Temple and perhaps also after its destruction, as we shall see later.

The statements of Josephus in *Antiquities* and in *Life*, quoted above, show that at the end of the Second Temple period it was customary to give the tithes to the priests.

While several passages in the Midrashim attest to the giving of the tithes to the priests or to both the priests and the Levites, it is difficult to fix the date of these passages and their relevance to actual con-

---

51 TB Soṭah 47b–48a; see also TJ Ma'aser Sheni v, 56d; TJ Soṭah ix, 24a: R. Jeremiah b. Ḥiyya in the name of Resh Laḳish.

52 Ch. Tchernowitz, "Demai — Hatza'ah Hisṭorit le-Taḳḳanot Yoḥanan Kohen Gadol she-ba-Mishnah," *Jewish Studies in Memory of G. A. Kohut*, New York, 1935, pp. 46–58; and *idem*, *Toledot ha-Halakhah*, New York, 1935–1950, III, pp. 135–148, holds that the ordinances of Johanan the high priest as a whole were meant to undermine the status of the Levites by doing away with their rights [transferring the first tithe to the priests, in consequence of which the confession that the tithe had been given was abolished and there was no longer any problem of demai]. Also abolished were the Levites' duties ["wakers" and "strikers" — the locking and opening of the Temple doors]. There is no discernible reason for associating the debased status of the Levites specifically with John Hyrcanus. The sources we have cited show that their diminished rights to the first tithe preceded the days of John Hyrcanus. Nor is there any reason to connect the "wakers" and "strikers" with the opening and closing of the doors, especially since the role of the Levites in this sphere is not clear (cf. Tamid i, 2). There is certainly no reason to emend the Mishnah by substituting "From his days" for "Until his days the hammer used to beat in Jerusalem" on the assumption that the beating of the hammer replaced the Levites' function of waking up the priests for their service.

ditions.[53] There is also the explicit difference of opinion on the question whether the first tithe was to be given to the Levite or to the priest. תנו רבנן, תרומה לכהן ומעשר ראשון ללוי דברי רבי עקיבא, רבי אלעזר בן עזריה אומר לכהן "Our Rabbis taught, Terumah belongs to the priest and the first tithe to the Levite, so R. Akiva. R. Eleazar b. Azariah said, To the priest."[54] In saying that the tithe was given to the priest, R. Eleazar b. Azariah expressed the view which accorded with the situation that existed during most of the Second Temple period, whereas R. Akiva upheld the theoretical halakhah which prescribed that the first tithe was given to the Levite.[55]

c) *The Tithes after the Destruction of the Second Temple*

Once the Temple was destroyed, the cessation of the priestly and levitical service in the Temple and the impossibility of bringing the

---

[53] Gen. Rabbah lxxi, 4, ed. Theodor-Albeck, p. 826; Ex. Rabbah xxxi, 17; Pesikta Rabbati xxv, ed. Friedmann, 126a.

[54] TB Yevamot 86a–b; TB Ketubbot 26a; TB Bava Batra 81b; TB Ḥullin 131b; and cf. TJ Ma'aser Sheni v, 56b–d; TJ Yevamot ix, 10b; x, 10d; TJ Soṭah ix, 24a. In the parallel passage in TB Bava Batra, R. Eleazar b. Azariah's view is given as מעשר ראשון אף לכהן "The first tithe belongs also to the priest." This is apparently not the original version, the word אף ["also"] having been transposed here from the continuation of the talmudic passage which reads, לכהן ולא ללוי אימא אף לכהן " 'To the priest,' but not to the Levite! — Read, To the priest also," that is, the Talmud questions R. Eleazar b. Azariah's view: How is it that the Levite has no tithe? To this the answer is given: Read, To the priest also. The view that the text in Bava Batra is corrupt is reinforced by the fact that the word אף does not occur in the Munich MS. where the Baraita is the same as in the other tractates. [In some MSS. and versions, the reading is ר׳ אלעזר בן עזריה אומר אף מעשר ראשון לכהן "R. Eleazar b. Azariah said, Also the first tithe belongs to the priest."]

[55] We cannot accept the view of Z. Carl, "The Tithe and the Terumah" (Hebrew), *Tarbiz,* 16 (1945), pp. 11–17, that after the destruction of the Second Temple the attempt was made to restore the biblical commandment which required that the first tithe be given to the Levite, since the service of the priests in the Temple had come to an end and the gifts were no longer vitally needed by their recipients. There is no parallel to the repeal of an enactment after the destruction of the Temple with the aim of reverting to the biblical law. What is more, the halakhah does not regard terumah and the tithes as a remuneration to the priests and the Levites for their Temple service, seeing that it lays down that these are to be given to any priest or Levite anywhere. Moreover the tendency of the halakhah after the destruction of the Temple was to preserve as far as possible the practices of the Second Temple period not only in the hope and belief that the Temple would speedily be rebuilt but also from the desire to fill the vacuum created among the people by the destruction of the Temple. Cf. also E. E. Urbach, "Social and Religious Trends in the Sages' Doctrine of Charity" (Hebrew), *Zion,* 16 (1951), pp. 11 ff.

tithes to Jerusalem and to the Temple as demanded by the ancient halakhah were enough, it seemed, to bring an end to the separation of the first tithe. And indeed an halakhah in this spirit occurs in the Jerusalem Talmud in tractate Sheḳalim. דתני אין מקדישין ולא מעריכין ולא מחרימין ולא מגביהין תרומות ומעשרות בזמן הזה, אם הקדיש או העריך או החרים או הגביה הכסות תישרף, הבהמה תיעקר, כיצד נועל בפניה הדלת והיא מתה מאיליה, והמעות ילכו לים המלח "It was taught, One may nowadays neither consecrate anything, nor make a valuation vow, nor declare anything devoted, nor separate terumot and tithes. If one has consecrated anything, or made a valuation vow, or declared anything devoted, or separated terumot and tithes, if it is a garment, it has to be burnt. If it is an animal, it has to be destroyed. How? The door is locked before it, so that it dies of itself. If it is money, it is cast into the Salt Sea."[56]

The commandments associated with the Temple and Jerusalem, such as blowing the shofar on the New Year that fell on a Sabbath, performing the ceremony of the lulav on all seven days of the festival of Tabernacles — these the Nasi and Sanhedrin at Jabneh wished to transfer to, and transform into precepts observed, wherever the Sanhedrin had its seat or anywhere, the aim being to fill the vacuum created by the destruction of the Temple.[57] This principle was also applied to the tithes in particular and to the other priestly gifts in general. The above halakhah, in so far as it reflects prevailing conditions, reflects only those that existed during the initial period following the destruction of the Second Temple. For it is clear that already during the subsequent period of Jabneh the separation of tithes and of other priestly gifts was reintroduced. This can be seen primarily from the numerous halakhot dealing with these topics and dating from after the destruction of the Temple. Nor can a merely theoretical significance be ascribed to all these halakhot, seeing that in several passages the Sages explicitly laid down that priestly and levitical gifts applied also after the destruction of the Second Temple even as during its existence.[58] This emerges also from the evidences furnished by various incidents, the most notable of which is the following: וכבר נשתהא רבי טרפון מלבא לבית המדרש, אמר

---

56 TJ Sheḳalim viii, 51b.

57 See, for example, G. Allon, *Toledot ha-Yehudim be-Eretz Yisra'el bi-Teḳufat ha-Mishnah ve-ha-Talmud*, Tel Aviv, I: 1958³, II: 1961² — see I, pp. 53–192; S. Safrai, "Beḥinot Ḥadashot li-Be'ayat Ma'amado u-Ma'asav shel Rabban Yoḥanan ben Zakkai le-aḥar ha-Ḥurban," *Sefer Zikkaron le-Allon*, Tel Aviv, 1970, pp. 203–226.

58 Sifrei, Ḳoraḥ, cxix, ed. Horovitz, p. 146; Sifrei Zuṭa, Ḳoraḥ, 18:21, ed. Horovitz, p. 297.

לו רבן גמליאל מה ראית להשתהות, אמר לו שהייתי עובד, אמר לו והלא כל דבריך תימה וכי יש עבודה עכשיו, אמר לו הרי הוא אומר "עבודת מתנה אתן את כהונתכם" – לעשות אכילת קדשים בגבולים כעבודת מקדש במקדש "Once R. Ṭarfon was late in coming to the bet ha-midrash. Rabban Gamaliel said to him, What is the reason for your late-coming? He answered him, I was engaged in the priestly service. He [Rabban Gamaliel] said to him, What you say is utterly surprising. Is there, then, any priestly service nowadays? He [R. Ṭarfon] answered him, It says, 'I give your priesthood as a service of gift,' thus making the eating of holy things outside the Temple and Jerusalem like the Temple service in the Temple."[59]

In the days of Bar Kokheva the laws relating to the tithes were widely observed, as can be seen from the leases signed by him. In them the tenants undertook to pay the landlord the amount of produce due to him, after the tithes had been separated.[60] The persecutions following the Bar Kokheva revolt included a prohibition against separating terumot and tithes,[61] which shows that their separation was widespread among the people.

It is important to point out that the Baraita cited above from tractate Sheḳalim in the Jerusalem Talmud does not, when quoted in the Babylonian Talmud, include the words ולא מגביהין תרומות ומעשרות "nor separate terumot and tithes,"[62] which undoubtedly indicates the reintroduction of the tithes in the period following the destruction of the Temple.

We have previously mentioned that after the Temple was destroyed there was a continuing difference of opinion whether the first tithe was to be given to the priests or to the Levites, the most noteworthy illustration of this being the argument between R. Akiva and R. Eleazar b. Azariah, discussed above. This difference of opinion assumed a practical character when, as we learn in the sequel, R. Akiva by a ruse prevented R. Eleazar b. Azariah, who was a priest, from getting the first tithe.[63]

---

[59] Sifrei, Ḳoraḥ, cxvi, ed. Horovitz, p. 133; TB Pesaḥim 72b–73a. Cf. also Sifrei Zuṭa, Ḳoraḥ, 18:7, ed. Horovitz, p. 293.

[60] See, for example, P. Benoit, O.P., J. T. Milik, R. de Vaux, O.P., *Les Grottes de Murabba'at* (*Discoveries in the Judaean Desert*, II), Oxford, 1961, No. 24, pp. 124–125.

[61] Ma'aser Sheni iv, 11; and see Allon, *Toledot ha-Yehudim*, II, pp. 45, 263.

[62] TB Yoma 66a; TB Bekhorot 53a; TB 'Avodah Zarah 13a.

[63] TB Yevamot 86b. See also Ma'aser Sheni v, 9; TJ Ma'aser Sheni v, 56b, in which there is a similar divergence of opinion between R. Joshua b. Hananiah and

Towards the end of the tannaitic and the beginning of the amoraic period the uncertainty continued whether the first tithe was to be given to the priests or to the Levites.[64] But a new element now entered into the separation of the tithes, in that they were given specifically to priests or Levites who were talmidei ḥakhamim, and even possibly in the course of time to talmidei ḥakhamim who were neither priests nor Levites. Here there obviously came into operation the tendency to give the priestly gifts to those who fulfilled public functions and bore the burden of the spiritual leadership of the nation, analogous to some extent to the role of the priests and the Levites in the days when the Temple existed.[65] An incident which reflects this situation occurs in the Jerusalem Talmud. רבי יונה יהב מעשרוי לר׳ אחא בר עולא, לא משום דהוה כהן, אלא משום

R. Eleazar b. Azariah. In Tosefta Pe'ah iv, 5 it is stated ר׳ שמעון בן גמליאל אומר, כשם שהתרומה חזקה לכהונה בחילוק גרנות, כך מעשר ראשון חזקה לכהונה בחילוק גרנות "R. Simeon b. Gamaliel [in the parallel passage in TB Ketubbot 26a: R. Simeon b. Eleazar] said, Even as the terumah is a presumption for the priesthood in the distribution of the priestly gifts of the threshing floor, so the first tithe is a presumption for the priesthood in the distribution of the priestly gifts of the threshing floor" (see also Tosefta Ketubbot iii, 1). This is the text according to the Erfurt MS., but in the printed version and in the Vienna MS. of Pe'ah and Ketubbot the reading is that the first tithe belongs to the Levites. This is apparently a later emendation, for in all the parallel passages the reading is as in the Erfurt MS. Cf. TB Ketubbot 26a; TJ Ketubbot ii, 26d; etc. Cf. also *Kaftor va-Peraḥ*, xxv, ed. Edelmann, 76a; and see Lieberman, *Tosefta ki-Peshuṭah*, Seder Zera'im, p. 181, who adopts the reading of the Vienna MS. and of the printed version.

64 R. Simeon b. Eleazar's statement in TB Ketubbot 26a, if we adopt this version in preference to that in the other parallel passages [see previous note]. As against this there are the statements: משנחשדו להיות נותנין מעשר לכהונה "Since they were suspected of giving the tithe to the priests" (TJ Soṭah ix, 24a; for the parallel passages, see note 54, above); and ביומוי דרבי יהושע בן לוי ביקשו להימנות שלא ליתן מעשר לכהונה "In the days of R. Joshua b. Levi they sought to take a vote [on the proposal] not to give the tithe to the priests" (TJ Ma'aser Sheni v, 56b).

65 So in Sifrei Numbers: "ונתתם ממנו את תרומת ה׳ לאהרן הכהן," מה אהרן חבר אף בניו חברים, מיכן אמרו אין נותנים מתנות אלא לחבר " 'And from it you shall give the Lord's offering to Aaron the priest' — Even as Aaron is a ḥaver, so are his sons ḥaverim. Accordingly they said, Priestly gifts are given only to a ḥaver" (Sifrei Numbers cxxi, ed. Horovitz, p. 148; in TB Sanhedrin 90b: in the name of דבי ר׳ ישמעאל תנא "The school of R. Ishmael taught..." Cf. also Mishnah Bikkurim in the name of R. Judah: אין נותנים אותם אלא לחבר בטובה "They are given only to [a priest who is] a ḥaver be-ṭovah": Bikkurim iii, 12; TJ Ḥallah iv, 60b. For the meaning of the expression ḥaver be-ṭovah, see S. Safrai, "Sabbatical Year Commandments under the Conditions Prevailing after the Destruction of the Second Temple" [Hebrew], *Tarbiz*, 35 [1966], pp. 314–316). "Ḥaver" is here used in the sense of a talmid ḥakham.

דהוי לעי באוריתא "R. Jonah gave his tithes to R. Aḥa b. 'Ulla, not because he was a priest, but that he might be able to study the Torah."[66] Mention should also be made of the statement of R. Abba bar Kahana in the Midrash: "עשר תעשר" — רמז לפרגמטופטין ולמפרשי ימים שיהו מוציאין אחד מעשרה לעמילי תורה " 'You shall tithe.' This is an intimation to merchants and seafarers that they are to give one-tenth to those who industriously study the Torah."[67] In the view of Allon, tithes were given to priests who were talmidei ḥakhamim as early as after the destruction of the Temple,[68] but this view is not fully substantiated.[69]

From a certain aspect it was specifically at a later period that the separation of tithes was extended to apply to other types of income, from which tithes were also to be separated.

The biblical laws pertaining to the tithes, and similarly the usages reflected in the Apocrypha, prescribed that tithes were to be separated from crops, wine, and oil.[70] The halakhah laid down the principle: כל שהוא אכל ונשמר וגדוליו מן הארץ חיב במעשרות "Whatever is considered food and is guarded and grows out of the soil is liable to tithes."[71] At a later stage, apparently at the end of the tannaitic period, the obligation to separate tithes was extended to include profits in general.

The Tosafists, in their comment on tractate Ta'anit,[72] quote an halakhic Midrash from the Sifrei which, extant neither in the printed versions nor in MSS., states "עשר תעשר את כל תבואת זרעך היוצא השדה שנה שנה," אין לי אלא תבואת זרעך שחייב במעשר, ריבית ופרקמטיא וכל שאר רווחים מנין, תלמוד לומר את כל, דהוה מצי למימר את תבואתך, מאי כל לרבות רבית ופרקמטיא וכל דבר שמרויח בו " 'You shall tithe all the yield of your seed, which comes forth from the field year by year.' From this we might infer that only 'the yield of your seed' is to be tithed. From what can we deduce that this applies also to interest, to business, and to all other profits? From the fact that the biblical passage says 'all,' since it could have said 'your yield.' Why, then, does it say 'all'? It is to include

66 TJ Ma'aser Sheni v, 56b.

67 Pesiḳta de-R. Kahana, chap. עשר תעשר, ed. Mandelbaum, p. 172 [פרגמטופטין and, in other versions, פרגמטוטין = merchants, from the Greek πραγματευτής].

68 Allon, *Toledot ha-Yehudim*, I, pp. 160–162.

69 In this connection, see Urbach, *Zion*, 16 (1951), pp. 10–13.

70 Lev. 27:30; Num. 18:27; Deut. 12:17; 14:23, 28; 26:12; etc.; Neh. 10:38; 13:5, 12; II Chron. 31:5. Judith 11:13; Tobit 1:6; etc. [For the tithe of cattle, see above, pp. 25–27.]

71 Ma'aserot i, 1; and see TB Nedarim 52a; cf. Tosafot TB Ro'sh ha-Shanah 12a, s.v. תנא.

72 Tosafot TB Ta'anit 9a, s.v. עשר תעשר.

interest, and business, and everything from which one makes a profit."[73]

Some evidence of the separation of tithes from goods and from money is to be found also in the source in the Pesiḳta de-R. Kahana quoted above — "this is an intimation to merchants and seafarers that they are to give one-tenth..."

The halakhot relating to priestly gifts, which occur in Didache and which, like the other halakhot in it, were presumably more or less faithfully copied from the prevailing usage in Israel, mention money, clothes, and chattels as things from which priestly gifts were separated.[74]

This wider separation of tithes was presumably due to the tendency to put an end to the difference between the farmer and the townsman in the matter of tithes, since the entire basis for the separation of tithes from agricultural produce only was rooted in an epoch when almost the whole nation were tillers of the soil. Yet it is evident that not all the people observed this wider separation of the tithes, which was rather in the nature of being "a pious custom" [מנהג חסידים].

As a "pious custom" we should also regard the procedure of separating tithes from all food, mentioned in the following statement in the Jerusalem Talmud: אמר רבי יוחנן כד הוה אכל אפילו קופד אפילו ביעה הוה מתקן "R. Johanan said, When I ate even a piece of meat, even an egg, I would separate the tithes."[75]

Concurrently with the desire to continue to observe the commandment of separating tithes and the Sages' insistence on it, there arose the tendency to introduce greater leniency into the law of the tithes in order to normalize economic conditions or to ease the people's burden at times of economic crisis. Thus after the destruction of the Second Temple instances of "evading" [הערמה] the law of the tithe increased, such as: ואמר רבה בר בר חנה אמר רבי יוחנן משום רבי יהודה בר׳ אלעאי בא וראה שלא כדורות הראשונים דורות האחרונים. דורות הראשונים היו מכניסין פירותיהן דרך טרקסמון כדי לחייבן במעשר, דורות האחרונים מכניסין פירותיהן דרך

73 Finkelstein contends that the statement of the Sifrei dates from the middle ages, for in the days of the Talmud the business of money-lending to non-Jews on interest had not yet spread among the Jews (Finkelstein in his edition of the Sifrei, pp. 165–166, note 15). There is no factual basis to Finkelstein's contention, since already in talmudic times Jews engaged in trading and in money-lending; and see Urbach, *Zion*, 16 (1951), pp. 12–13.

74 Didache 13:3. And see Allon, *Meḥḳarim*, I, pp. 292–294.

75 TJ Demai i, 21d. The explanation of the Talmud דו חשש למשקין שיש בהן — that it is "because he was concerned about the liquids in them" cannot of course be reconciled with the literal meaning.

גגות דרך חצרות דרך קרפיפות כדי לפטרן מן המעשר "Rabbah bar bar Ḥanah said in the name of R. Johanan reporting R. Judah b. Ila'i, Come and see the difference between the earlier and the later generations. The earlier generations used purposely to bring in their produce by way of the kitchen-garden in order to make it liable to the tithe, whereas the later generations bring in their produce by way of roofs or backyards or enclosures in order to exempt it from the tithe."[76] Another example of "evading" the law of the tithe is the following: תני ר׳ הושעיה מרבה אדם דגן בתבן, ומערים עליו לפוטרו מן המעשרות "R. Hoshaia taught, A man increases grain in the straw, thus resorting to a device to exempt it from the tithes."[77] There were enactments which annulled the obligation to separate tithes in border regions, for not only was it doubtful whether the commandments associated with the produce of the land in Eretz Israel applied to them, but there was also the fear of the border regions' complete estrangement because of the sparse Jewish settlement in them. Here there accordingly entered a further tendency towards leniency in the laws relating to the tithes due to the desire to enlarge and consolidate Jewish settlement. In this connection special mention should be made of the halakhot and reforms of R. Judah ha-Nasi,[78] similar to his reforms relating to the Sabbatical Year and the purchase of confiscated property [סיקריקון], the general tendency being to bring about a normalization in the nation's economy and settlement.[79]

A further relief in the laws of the tithes took place in the period of anarchy in the third century CE, when the prevailing economic crisis created a need for leniency as regards the tithes.[80] The view that the

---

[76] TB Berakhot 35b; TB Giṭṭin 81a; and see TJ Ma'aserot iii, 50c.

[77] TJ Berakhot v, 8d; and cf. TB Berakhot 31a; TB Pesaḥim 9a; TB 'Avodah Zarah 41b; TB Menaḥot 67b; TB Niddah 15b.

[78] TJ Demai ii, 22c; TB Ḥullin 6b; TJ Shevi'it vi, 36c. In one of the sources it is clearly stated that border regions were exempt from tithes and from the Sabbatical Year but אין בהן משום ארץ העמים "they do not come into the category of heathen countries" (Tosefta Ahilot xviii, 4). Hence these regions were defined as foreign countries falling outside Eretz Israel only as far as tithes and the Sabbatical Year are concerned.

[79] See Allon, *Toledot ha-Yehudim*, II, pp. 153–158; S. Safrai, "Siḳariḳon," *Zion*, 17 (1952), pp. 56–64; *idem*, *Tarbiz*, 36 (1967), pp. 26–37.

[80] We find the association דבי רבי ינאי "of the school of R. Yannai," which stayed at 'Akhbarah during the period of anarchy, and which imposed upon itself restrictions in various matters after the manner of associations and sects, nevertheless adopting a lenient attitude to tithes: אישתאלית לאילין דבי רבי ינאי, ואמרין נהגין הוינן יהבין אילין לאילין בחקלא ואכלין ולא מתקנין "I asked members of

tithes were now no longer biblically prescribed, but מאיליהן קיבלו "were an obligation voluntarily assumed,"[81] was also undoubtedly put forward to serve as a theoretical background for introducing leniency into the laws of the tithes. Yet from time to time we come across expressions of opposition on the part of the Sages to such lenient reforms of the laws of the tithes,[82] thus once more manifesting the extent to which the Sages ascribed importance to the subject.

d) *The Separation of Tithes in the Diaspora*

According to the theoretical halakhah, terumot and tithes are not practised outside Eretz Israel. Every precept, which is an obligation arising from the soil, is considered a commandment associated with the produce of the land in Eretz Israel, and is thus not applicable outside the country. This principle naturally also refers to terumot and tithes, as well as to other priestly gifts.[83] In addition to this, numerous halakhic sources explicitly lay down that tithes do not apply to the diaspora.[84]

And yet many sources testify that there were in fact people who separated tithes outside Eretz Israel. It is mainly with Babylonia that the talmudic sources naturally deal,[85] but from them we can nevertheless infer what the situation was in other countries of the diaspora.[86]

The sources in the halakhah attest not only to the separation of tithes in the diaspora but also to the fact that they were brought to Eretz Israel. This means apparently that the tithes were not given to priests or Levites living in the diaspora but were sent to Eretz Israel,[87] most probably on the occasion of a pilgrimage.

Several incidents reported at the end of Mishnah Ḥallah testify to the custom of bringing priestly gifts from the diaspora to Jerusalem. True, no reference is made there to the tithe, and it is also stated there

the school of R. Yannai, and they said, We were accustomed to give to one another [from the produce] in the field and to eat and did not separate the tithes" (TJ Ma'aserot ii, 49c).

81 See above, note 35 (p. 37).

82 TJ Demai ii, 22c; TB Ḥullin 6b.

83 Ḳiddushin i, 9. On the antiquity of Ḳiddushin chap. i, see J. N. Epstein, *Mevo'ot le-Sifrut ha-Tanna'im*, Jerusalem, 1957, pp. 52–54.

84 Ḥallah ii, 2; TJ Berakhot iv, 7b; TB Shabbat 119a.

85 See, for example, Avot de-R. Nathan, Version A, xx, ed. Schechter, p. 73.

86 We shall not here go into detail about the regions concerning which it was doubtful whether they belonged to Eretz Israel or to a foreign country.

87 Yadayim iv, 3.

that לא קבלו "they did not accept" these gifts from them. Yet the very mention of such incidents shows that there was the custom of bringing priestly gifts from the diaspora to Jerusalem, including undoubtedly also terumot and tithes.[88] Moreover many halakhot in Mishnah Ḥallah, among them some that have been mentioned, were the subject of conflicting opinions, so that there was apparently also a view stating that priestly gifts were to be accepted from the diaspora.[89]

In the halakhah pertaining to the tithe of the cattle there was similarly a difference of opinion whether it was to be brought from the diaspora or not.[90]

Josephus mentions an edict issued by the pro-consul of Asia Minor to the people of Miletus in the days of Caesar, charging them to permit the Jews to "manage their produce in accordance with their custom."[91] Philo admonished those who refrained from separating terumot and tithes, and in doing so very likely referred to his fellow-Jews in Alexandria and in Egypt.[92] The Church Father Epiphanius tells that Joseph the apostate, before his conversion to Christianity, acted as a "delegate of the Nasi," collecting from each city the tithes.[93]

Further proof that tithes were brought from the diaspora to Eretz Israel may be found in the evidences about the sum, known as the aparche (ἀπαρχή), which the Jews had to pay as an addition to the tax of two denars. After the destruction of the Second Temple the Jews both in Eretz Israel and in the diaspora were compelled to pay to the Fiscus Judaicus in Rome a tax of two denars in place of the half-shekel.[94] Potsherds from Edfu and papyri from Arsinoë in Egypt show that to the sum of two denars, which replaced the half-shekel, there was added the further sum of a drachm, which was called the aparche. From the papyri of Arsinoë we learn that it was levied as early as in 72, 73 CE, while on the Edfu potsherds it is mentioned from Vespasian's fifth year. The term aparche is apparently to be translated here as

---

[88] Ḥallah iv, 10–11: see Epstein, *Mevo'ot le-Sifrut ha-Tanna'im*, p. 273.

[89] Temurah iii, 5; and see TB Temurah 21a–b; cf. Sifrei, Re'eh, cvi, ed. Finkelstein, p. 166; and Tosefta Sanhedrin iii, 5. Several halakhot in Mishnah Ḥallah, cited above, were also the subject of conflicting views.

[90] Bekhorot ix, 1; and TB Bekhorot 53a.

[91] Josephus, *AJ*, xiv, 10, 21, §§244 ff.

[92] Philo, *De Specialibus Legibus*, I, §§153–155.

[93] This levy was apparently assigned for the needs of the Sages of Eretz Israel, a development we have dealt with above, pp. 45–46.

[94] See Allon, *Toledot ha-Yehudim*, I, pp. 39–41, and the bibliography *ad loc.*

"terumah," this addition to the tax being probably instead of the terumot and the tithes which the Jews of the diaspora used to bring to Eretz Israel.[95]

From the meagre value of this sum we may infer that the Jews of the diaspora did not bring the full amount of the terumot and the tithes to Jerusalem, but rather satisfied themselves with redeeming them for less than their actual value, and brought that sum to Jerusalem.[96]

Presumably at a certain stage, as also happened in Eretz Israel, the terumot and tithes from abroad were converted into a tax which was given to the Sages for their maintenance instead of to the priests and the Levites. This accords with the later evidences of the Church Fathers, one of which we cited above, and which deal with the institution of the "delegates of the Nasi" who went out to collect these "terumot and tithes." This custom was apparently not instituted in the days of the Church Fathers but had its origin in earlier times, either in the period of the Severans or perhaps even before that. Possibly connected with it is the information about מגבת חכמים "the collection of contributions for the maintenance of Sages," which may have been seized upon by the Jews of the diaspora as a substitute for the terumot and the tithes.

B. PURITY AND IMPURITY

1. Purity and Impurity in the Bible

Biblical law states that the body of a human being and of an animal, plants, and inorganic substances are liable to become impure. The impurity can reside in the impure object itself or can be conveyed from

[95] See V. Tcherikover, *Ha-Yehudim be-Mitzrayim ba-Teḳufah ha-Hellenisṭit Romit le-Or ha-Papyrologyah,* Jerusalem, 1963³, p. 92, and the bibliography *ad loc.* The view of Blau (see Tcherikover, *op. cit., loc. cit.*) that the aparche replaced the first fruits (i.e. aparche is to be translated here as "first fruits") is improbable, since there is no proof that at any time first fruits were brought from the diaspora [there was even a dispute among the Tannaim whether it was necessary to bring first fruits from Transjordan: see Bikkurim i, 10].

[96] The aparche tax is mentioned only in Egyptian documents, but since it was combined with the payment of the half-shekel which was imposed upon the whole Jewish diaspora in the Roman Empire, it too may have been laid on all the members of the Jewish diaspora in the Roman Empire. If this is so, it is further proof that tithes were separated by the diaspora in the days of the Second Temple.

without by contact, by carriage, and, in the case of a corpse, also by overshadowing, that is, by being under the same roof with the dead body.

The principal causes of impurity are:

i) Leprosy — the leprosy of a human being, of clothes, and of houses.

ii) Secretions from the sexual organs — as in the case of a menstruating woman, of a man or a woman with a discharge from the flesh, of one who has had a nocturnal seminal emission, of a husband and wife after sexual intercourse, and of a woman after childbirth.

iii) Dead bodies — a human corpse, an animal carcass [except in the case of a clean, permitted animal slaughtered according to Jewish law], and the bodies of the eight reptiles mentioned in the Bible as impure when dead.

These causes of impurity differ from one another in their severity, in the process and manner of purification from them, and in the interval of time that has to elapse until purification takes place.[97]

In the Bible the term impurity is also applied to prohibited foods, to unchastity, and to idolatry.

Outside the biblical legal chapters, the Bible ascribes impurity to other things too — to an uncircumcised person, to foreign territory, and to idols.[98] [99]

2. The Relation of Purity and Impurity to Things Sacred and Secular

In matters relating to purity and impurity the Bible reveals two different approaches. Thus there are verses that deal with the laws of purity

---

97 Lev. 5:2–3, 11–17; Num. 19; 31:19–20; Deut. 14:3–21; 23:10–15; 24:7; 26:14; etc.

98 The uncircumcised — Is. 52:1; etc.; foreign territory — Jos. 22:19; Hos. 9:3; Amos 7:17; idols — Gen. 35:2; Is. 30:22. The prophets also used the term impurity metaphorically for idolatry and bloodshed. In general it seems that according to the prophetic outlook every sin engenders impurity, and its atonement produces purity.

99 On the subject of purity and impurity in the Bible, see Kaufmann, *Toledot ha-Emunah ha-Yisra'elit*, I, pp. 539–559; J. Döller, *Die Reinheits- und Speisegesetze des Alten Testaments*, Münster, 1917; J. Scheftelowitz, "Das Opfer der roten Kuh (Num. 19)," *ZAW*, 39 (1921), pp. 113–123; W. H. Gipsen, "Clean and Unclean," *OTS*, 5 (1948), pp. 190–196; L. Koehler, "Aussatz," *ZAW*, 67 (1955), pp. 290–291; P. Reymond, "L'eau, sa vie et sa signification dans l'Ancien Testament," Supplement VI to *VT*, 1958.

and impurity solely in relation to priests, to the eating of sacred things, and to entering the Temple,[100] and, on the other hand, there are also verses which, according to their literal meaning, deal with the laws of purity and impurity in relation to all Israel everywhere, quite unconnected with the Temple and with sacred things.[101] [102]

In its application of the concepts of purity and impurity to sacred and to secular things, the halakhah adopts both these approaches.

The theoretical halakhah lays down that the laws of purity and impurity refer only to the priests, to the Temple, and to the eating of sacred things, and it is in accordance with this view that the halakhah tends to interpret and explain the biblical verses that deal with the laws of purity and impurity relating to all Israel everywhere.[103]

On the other hand, talmudic literature contains statements which enlarge the application of the laws of purity and impurity beyond the

---

[100] Lev. 7:19–21; 12:4; 22:4–8; Num. 19:20.

[101] Lev. 11:8; etc.

[102] A similar dualism occurs in the Bible in connection with making a baldness on the head, making cuttings in the flesh, and shaving off the edges of the beard, which refer on one occasion to priests, on another to all Israel (Lev. 21:5 as against Lev. 19:27–28 and Deut. 14:1; etc.).

[103] The verse in Lev. 11:8, which according to its literal meaning states that the impurity of a carcass refers to all Israel, is interpreted in the Sifra as follows: "ובנבלתם לא תגעו", יכול יהו ישראל מוזהרים על מגע נבילות, ת״ל אמור אל הכהנים בני אהרן ואמרת אליהם לנפש לא יטמא בעמיו, הכהנים אין מטמאים למתים, ישראל מטמאים למתים " 'And their carcasses you shall not touch.' I might think that all Israel are cautioned not to touch carcasses. Therefore the Bible says, 'Speak to the priests, the sons of Aaron, and say to them that none of them shall defile himself for the dead among his people.' The priests do not defile themselves for the dead, Israelites do defile themselves for the dead" (Sifra, Shemini, ii, ed. Weiss, 49a). See also TB Ro'sh ha-Shanah 16b, where the verse in Lev. 11:8 is explained as applicable only on festivals.

The text in Lev. 14:36 deals with the obligation to keep vessels from becoming impure, and according to its literal meaning refers to all Israel. R. Meir interprets this text as indicating that the Torah has consideration for the property of Israel, that they should not suffer a loss, since the vessels would be unfit for the preparation of food in ritual purity. According to R. Judah, the vessels have to be taken out of the house in which a sign of leprosy has appeared since it is so laid down in the text and not because of impurity (Nega'im xii, 5; Sifra, Metzora', v, ed. Weiss, 73a. And see Allon, *Meḥḳarim*, I, p. 148, note 3).

The injunction to send the impure out of the camp is explained in the halakhah as referring, within the area of Jerusalem, only to the Temple Mount. Lepers too are, according to the halakhah, sent out of walled cities only, and are permitted to go about throughout the country (Kelim i, 7–8; Sifrei Zuṭa, Naso, 5:2, ed. Horovitz, p. 228).

limits of the Temple precincts, to objects other than sacred things, and to all Israel, even to the extent of demanding that purity be observed everywhere and at all times, in order that all Israel shall attain a sanctity like that of the priests and even secular food shall be eaten in a state of ritual purity.[104]

The obligation to observe purity everywhere and at all times is evident also from statements of Philo and Josephus who in various passages mention the duty of one who had contracted corpse-impurity to purify himself, although there was no question of his going to the Temple.[105] They also stress that anyone impure with corpse, menstruation, or other impurities had to keep away from ritually pure persons and things to preclude their contracting impurity.[106] Ritually pure persons had, therefore, to refrain at all times from contact with impurity, even without reference to sacred things or to the Temple.

These divergent views on the extent to which the laws of purity and impurity were to be applied occupied a central place in the differences of opinion and of approach which characterized the parties and sects in Israel in the days of the Second Temple. The Sadducees took the view which limited purity and impurity to the priests, the Temple, and sacred things,[107] while the Essenes, the members of the Dead Sea sect, and the ḥaverim adopted the approach which applied the laws of purity and impurity to all times and to everyone in Israel. The Pharisees wavered between, on the one hand, the theoretical acceptance of the extended application of purity and impurity and, on the other, the adoption of a lenient view in instances in which the public were unable

---

104 A number of halakhic sources indicate, for example, that one who has had a nocturnal emission of semen [בעל קרי] has on each occasion to immerse himself in a ritual bath to purify himself from the impurity (TJ Berakhot iii, 6c; TB Berakhot 22a; TB Yoma 88a; etc.). Several sources give the halakhot relating to the purity of secular food without distinction between those who eat their secular food in a state of ritual purity and those who do not (Ḥullin ii, 5; Tosefta Berakhot vi, 2–3; Tosefta Makhshirin iii, 7). On the subject of eating secular food in a state of ritual purity, see later. On the view that all Israel should be pure like priests, see Seder Eliyahu Rabbah xvi, ed. Friedmann, p. 72.

105 Philo, *De Specialibus Legibus*, III, §§205–206; Josephus, *Against Apion*, ii, 23, §198; 26, §205.

106 Philo, *op. cit.*, III, §63; Josephus, *AJ*, iii, 11, 3, §§261 ff. (and cf. Seder Eliyahu Rabbah xvi, ed. Friedmann, pp. 75–76).

107 This view of the Sadducees arose primarily from the fact that they constituted the majority of the priesthood and its principal strength. See Allon, *Meḥḳarim*, I, pp. 148–176; R. Leszynsky, *Die Sadduzäer*, Berlin, 1912, pp. 280 ff.

in their daily life to endure some restriction. This duality on the part of the Pharisees explains the existence of the two approaches in Pharisaic law.[108]

3. The Place of the Laws of Purity and Impurity in the Halakhah

The laws of purity and impurity occupy an extensive place in talmudic literature. An entire Order, that of Ṭohorot, consisting of twelve tractates, is devoted to the subject, in addition to which talmudic literature comprises Mishnayot, Baraitot, halakhic Midrashim, and countless discourses [sugyot] that deal with the subject.

Even as there is a similarity between the biblical laws of purity and impurity and those of the ancient east — of Egypt, Mesopotamia, Babylonia, and other lands — so a relation may be found to exist between the relevant talmudic halakhah and the laws of purity and impurity among other nations, this being particularly so in the case of the customs associated with purity and impurity as practised in the temple cults of Greece.

The considerable extent of the laws of purity and impurity in talmudic literature alongside other evidences shows the great importance ascribed to these laws and the central, prominent place occupied by them in daily life.

The talmudic halakhah demands that purity be observed in the realm of the body, of clothes and vessels, and of food.

i) *The Body*. The Bible states that a man contracts impurity from a corpse in one of two ways: either by contact with it (Num. 19:11) or by overshadowing, that is, by being under the same roof-space with it (Num. 19:12–14). To these two, the halakhah has added a third way of contracting impurity from a corpse, and that is by carrying it (Sifrei, Ḥuḳḳat, cxxvii, ed. Horovitz, p. 164). A person rendered impure by a corpse in one of these ways becomes "a father of impurity" [אב הטומאה], that is, a generating, principal impurity. Accordingly, the halakhah defines a corpse, which is the highest degree of impurity, as "the father of fathers of impurity" [אבי אבות הטומאה], in contrast to the other impurities mentioned in the Torah, which are regarded as "fathers of impurity."

If a person has, for example, a discharge or leprosy, he himself becomes a source of impurity in the degree of "a father of impurity." One ren-

---

[108] See Allon, *op. cit.*, I, pp. 175–176.

dered impure by "a father of impurity" becomes "an offspring of impurity" [ולד הטומאה], that is, a generated, secondary impurity of the first degree. A person does not contract impurity from "an offspring of impurity."

One ritually impure is purified by immersion in a ritual bath [מקוה], except in the case of impurity caused by a discharge when purification is by immersion in a flowing spring. While the time that has to elapse between the moment of becoming impure and immersion in a ritual bath varies according to the impurity and its degree, in every instance an impure person who has taken a ritual bath does not become pure until the end of the day of the immersion, that is, until sunset. During the intervening period he is defined as one who has immersed himself but must wait until sunset to become completely pure [טבול יום], and during this period he renders terumah and holy things "unfit" [פסול]. [a] One who has contracted corpse-impurity has to be sprinkled on the third and seventh days of his impurity with water of purification [מי חטאת: water sanctified with the ashes of the red heifer]. After the second sprinkling he immerses himself in a ritual bath and is impure until the evening.

Biblical law lays down that a person's impurity affects his body in its entirety, there being no impurity of individual limbs. As previously mentioned, a person contracts impurity from "a father of impurity" and is rendered "an offspring of impurity," becoming impure in the first degree [ראשון לטומאה], but does not contract impurity from "an offspring of impurity," so that should he come in contact with "an offspring of impurity" he continues to be ritually pure.

However the halakhah declares that there can be impurity of the hands alone while the rest of the body remains pure. Hence if the hands come in contact with an impurity of the first degree, they become impure in the second degree [שני לטומאה]. They also contract impurity when they touch the Scriptures or tefillin. This is an ancient ordinance which apparently goes back to the days of the Second Temple and which according to one talmudic tradition was instituted by Hillel and Shammai,[109] according to another by king Solomon,[110] ascriptions which attest to the antiquity of the ordinance.

---

109 TB Shabbat 15a.

110 TB 'Eruvin 21b.

a) At this stage no further impurity is engendered and the series of impurities comes to an end.

At a later stage it was laid down that in every case there is a presumption that hands are impure, touching things, as they do, automatically, so that there is the constant fear that they came in contact with "an offspring of impurity." The halakhah therefore states that even if a man is ritually pure, should his hands, about which there is the presumption that they are impure unless they were previously washed, come in contact with terumah, they make it unfit [since a second degree of impurity renders terumah unfit].[111]

ii) *Vessels and Clothes.* The Bible mentions clothes, vessels of wood and of skin, and sacks as susceptible to impurity, all these being rendered impure by contact. While earthen vessels are liable to contract impurity, they only defile if the source of the impurity falls into them (Lev. 11:32–35; etc.).

From the Bible the halakhah inferred that bone vessels are susceptible to impurity,[112] unlike vessels of dung, stone, or clay, which do not contract impurity.

An enactment of considerable antiquity, reported in the names of Jose b. Joezer and Jose b. Johanan [the first Pair after Simeon and Antigonus; Jose b. Joezer lived before the Hasmonaean revolt], decrees that glass vessels, too, are susceptible to impurity.[113]

The transference of impurity from one vessel to another, or from a vessel to a person or to food and the manner of purification from the impurity are similar to what has been said above in connection with the impurity of the body.

iii) *Eatables.* Food and liquids contract impurity, the significance and extent of which will be dealt with later. Food is rendered impure only if it is unattached to the ground and moistened by one of the seven liquids that make it liable to impurity.[114] When food becomes impure, the impurity is transferred to anything attached to it.

---

[111] Those scrupulous about eating secular food prepared in ritual purity are enjoined to wash their hands; see later.

[112] Sifra, Shemini, ix, ed. Weiss, 56a; TB Ḥullin 25b.

[113] TJ Shabbat i, 3d; TB Shabbat 14b; and cf. Tosefta Kelim, Bava Batra, vii, 7. There was an economic reason for this enactment: many preferred to use imported glass vessels which were not susceptible to impurity rather than earthen and metal ones made in Eretz Israel. The enactment may also be part of the struggle against Hellenistic culture, which was associated with glass articles and their spread (see L. Ginzberg, *'Al Halakhah ve-Aggadah*, Tel Aviv, 1960, pp. 14–15).

[114] The seven liquids that render produce liable to impurity are wine, blood, oil, milk, dew, honey, and water (Makhshirin vi, 4). The halakhah also mentions liquids which are subspecies of water, and deals with the liquids which cause both impurity

Food cannot become "a father of impurity," so that even if it comes in contact with a corpse, it is made "an offspring of impurity" in the first degree. Accordingly food does not render a man or a vessel impure, as these contract impurity solely from "a father of impurity" [food makes a person impure only if he eats it; see later].

4. Degrees of Purity and Impurity in Food

The observance of ritual purity in food acquired a special dimension and significance in daily life by reason of the custom adopted by sects and by certain circles of maintaining the ritual purity of their food at all times and of eating even their secular meals in ritual purity. This of necessity led them, for fear of transgressing through impure food, to erect social barriers between themselves and those in the nation who were not so scrupulous in observing the purity of food.

Several times the halakhah mentions the obligation of immersion in a ritual bath in order that food may be eaten in purity. So, for example, the Mishnah refers to the taking of a ritual bath by certain categories of persons prior to their eating the paschal lamb.[115] The need of one who has had a nocturnal seminal emission to immerse himself in a ritual bath is also for the purpose of partaking of food: אמר רבי יעקב בר אבון, כל עצמן לא התקינו את הטבילה הזאת, אלא שלא יהו ישראל כתרנגולין הללו משמש מיטתו ועולה ויורד ואוכל "R. Jacob b. Abbun said, The actual reason for their instituting this immersion in a ritual bath is that Israel should not be like these cocks, copulating, getting up, going down, and eating."[116]

A Mishnah compiled in the days of the Second Temple reflects the degrees of purity in food. הטובל לחלין והחזק לחלין, אסור למעשר. טבל למעשר והחזק למעשר, אסור לתרומה. טבל לתרומה והחזק לתרומה, אסור לקדש "One who immersed himself in a ritual bath for the purpose of eating secular food and intended to be rendered fit solely for secular food is prohibited from eating the tithe. One who immersed himself in a ritual bath for the purpose of eating the tithe and intended to be rendered fit solely for the tithe is prohibited from eating terumah. One who

---

and susceptibility to impurity, as also with those which cause only impurity but not susceptibility to impurity.

[115] Pesaḥim viii, 8.

[116] TJ Berakhot iii, 6c; and see Allon, *Meḥḳarim*, I, p. 150.

immersed himself for the purpose of eating terumah and intended to render himself fit solely for terumah is prohibited from eating sacred things."[117] Some sources have no special degree for the tithe, whose degree is the same as that for secular food, so that the tithe in the Mishnah may refer to the second tithe; in other sources it has the same degree as terumah. The degrees of purity in food are therefore as follows: i) sacred things [the parts of the sacrifices eaten by the priests]; ii) terumah [eaten by the priests; corresponding to it are certain kinds of sacrifices and the second tithe which were eaten by all Israel]; iii) secular food [other eatables].

It is therefore evident that the laws of purity can be observed at various degrees. Whoever observes the laws of purity in food intends a certain degree of purity and keeps the rules prescribed for that degree. Hence if he immerses himself in a ritual bath in order to become purified, he does so to attain a specific degree of purity. Something of this is reflected in the Mishnah. — עשר טמאות פורשות מן האדם: מחסר כפורים אסור בקדש, ומתר בתרומה ובמעשר. חזר להיות טבול יום – אסור בקדש ובתרומה, ומתר במעשר. חזר להיות בעל קרי – אסור בשלשתן... "Ten grades of impurity emanate from a man. Prior to the offering of his obligatory sacrifices, he is forbidden to eat holy things but is permitted terumah and the tithe. If he has immersed himself in a ritual bath but has still to wait until sunset to be completely pure, he is forbidden to eat holy things and terumah but is permitted the tithe. If he has a nocturnal seminal emission, he is forbidden to eat any of the three..."[118]

Of those who ate secular food in a state of ritual purity some were scrupulous in eating it at the degree of the purity of terumah or even of holy things. This led to the formulation of the concepts mentioned in the halakhah: חולין שנעשו על טהרת התרומה "Secular food prepared at the degree of ritual purity required for terumah," and חולין שנעשו על טהרת הקודש, "Secular food prepared at the degree of ritual purity required for holy things." These special degrees which were observed by individuals and by certain sects are expressed in the Mishnah. בגדי עם הארץ מדרס לפרושין. בגדי פרושין מדרס לאוכלי תרומה. בגדי אוכלי תרומה מדרס לקדש. בגדי קדש מדרס לחטאת. יוסף בן יועזר היה חסיד שבכהנה והיתה מטפחתו מדרס לקדש. יוחנן בן גדגדא היה אוכל על טהרת הקדש כל ימיו, והיתה מטפחתו מדרס לחטאת "The garments of an 'am ha-aretz are a source of

---

[117] Ḥagigah ii, 6. On the antiquity of Mishnah Ḥagigah in general and of this Mishnah in particular, see Epstein, *Mevo'ot le-Sifrut ha-Tanna'im*, pp. 46–52.

[118] Kelim i, 5.

midras-impurity[a] to Pharisees, the garments of Pharisees are a source of midras-impurity to those who eat terumah, the garments of those who eat terumah are a source of midras-impurity to those who eat holy things. The garments of those who eat holy things are a source of midras-impurity to those who attend to the water of purification. Jose b. Joezer was the most pious of the priests and his apron was a source of midras-impurity to those who ate holy things. All his life Johanan b. Gudgada ate secular food at the degree of ritual purity required for holy things, and his apron was a source of midras-impurity to those who attended to the water of purification."[119]

Food which comes in contact with a source of impurity, including other impure food, is rendered impure if it is susceptible to impurity. With every transference of the impurity, its severity diminishes by one degree. The number of times that an impurity is transferable depends on the degree of the purity of the food that comes in contact with the source of the impurity. For the higher the degree of the purity of the food, the more numerous the possible transferences of the impurity, as set out in Mishnah Ṭohorot.[120]

Liquids, which are an exception to all this since they are marked by a greater severity, contract impurity in the first degree even when defiled by an impurity in the second degree. This circumstance is based on the principle that משקין תחילה לעולם "liquids are always rendered impure in the first degree." Hence liquids which come in contact with an impurity of the second degree can render even secular food moistened by them impure in the second degree.[121]

---

[119] Ḥagigah ii, 7. This Mishnah is likewise extremely ancient. One of the proofs that it dates from the days of the Second Temple derives from the fact that the parallel passage in the Tosefta has the addition רבן גמליאל היה אוכל על טהרת חולין כל ימיו... אונקלוס הגר היה אוכל על טהרת הקודש כל ימיו "All his days Rabban Gamaliel ate secular food prepared in ritual purity... All his days Onḳelos the proselyte ate secular food prepared according to the degree of purity required for sacred things" (Tosefta Ḥagigah iii, 2–3). Those mentioned in the Mishnah as scrupulous in observing ritual purity lived in the days of the Second Temple, while the Tosefta adds the names of those who belonged to the period of Jabneh. The Mishnah was accordingly compiled in the days of the Second Temple. [On the contents of the halakhot in the Tosefta, see later.]

[120] Ṭohorot ii, 2–6.

[121] Tosefta Ṭevul Yom i, 3. In i, 6 mention is made of a further restriction which

a) Impurity contracted by an object on which one with a discharge (see Lev. 12:2; 15:2, 25) sits, treads, lies, or leans [pressure-impurity]. The impurity is conveyed to anyone who carries, or is carried on, the object.

If a man eats impure food he himself becomes impure.[122] The degree of his impurity is the subject of divergent views. רבי אליעזר אומר, האוכל אכל ראשון – ראשון, אכל שני – שני, אכל שלישי – שלישי. רבי יהושע אומר, האוכל אכל ראשון ואכל שני – שני, שלישי – שני לקדש, ולא שני לתרומה. בחלין שנעשו לטהרת תרומה "R. Eliezer said, He who eats food of the first degree of impurity is rendered impure in the first degree; of the second degree, is rendered impure in the second degree; of the third degree, is rendered impure in the third degree. R. Joshua said, He who eats food of the first or second degree of impurity is rendered impure in the second degree; of the third degree, is rendered impure in the second degree with regard to holy things but not with regard to terumah. All this applies to secular food prepared at the degree of ritual purity required for terumah."[123]

5. The Social Implications of the Laws of Purity and Impurity

The central position occupied by the laws of purity and impurity in the religious consciousness of the Jews in the days of the Second Temple, the profusion of these laws, and the difficulty of observing them led to the creation of social-religious barriers. Thus a barrier was created between, on the one hand, individuals, classes, and sects that were scrupulous in observing the laws of purity and impurity, and sometimes even added further restrictions to them, and, on the other hand, those who were not quite so scrupulous and at times even displayed a disregard for these laws.

At one extremity of this social scale were the members of the Dead Sea sect, as also the Essenes, and the ḥaverim, whose life and daily activities centred round the observance of the restrictions of purity and impurity; at the other extremity were the ʻammei ha-aretz who were suspected of not observing and of disregarding the laws of purity and impurity.

Between these two extremities we find a wide spectrum of individuals and classes that were in various degrees scrupulous in observing the laws of purity and impurity. Examples of this segmentation are to be

---

applies to liquids in relation to food, namely, that the least amount of water conveys impurity.

122 Zavim v, 12.

123 Ṭohorot ii, 2.

found in the early Mishnah in Ḥagigah (ii, 7), which we have quoted above, and which mentions the ʿammei ha-aretz and the Pharisees, those who ate terumah and those who ate holy things, as well as individuals who imposed greater restrictions upon themselves in various degrees of the scrupulous observance of purity and impurity. The consequence of this segmentation was the creation of barriers between one class and another, as set forth in the Mishnah which declares that the garments of each degree are a source of midras-impurity to the degree above it. The term "midras" means that the impurity of their garments is identical with that of the articles or garments on which one with a discharge treads, sits, lies, or leans. The practical significance of this is that these garments are "a father of impurity," rendering a person impure by contact and by carriage, and an article impure by contact. Thus contact between those scrupulous and those not scrupulous in observing the laws of purity and impurity was restricted by these very laws themselves. For clearly someone eating, for example, secular food in a state of ritual purity was unable to buy his food from anyone who did not scrupulously observe purity.

While the Sages were generally scrupulous in matters relating to purity and impurity, there were among them also various degrees ranging from an excessively scrupulous observance to a certain disregard of these laws. Thus, for instance, some Sages, who belonged to the associations of the ḥaverim, were so scrupulous in observing the laws of purity and impurity that they imposed additional restrictions upon themselves, whereas other Sages, who belonged to the ḥasidim and were held up as an example by reason of their character and their social qualities, were it seems not inordinately strict in matters of purity and impurity.[124] There is also evidence that Sages were opposed to the excessively scrupulous observance by priests of the laws of purity and impurity.[125]

Opposition on principle to the strict observance of the laws of purity came from the Christians, for these laws undoubtedly played an important role in bringing about their segregation from the normative community. Antagonism to and scorn of the Pharisees' observance of the laws of purity and impurity find vigorous expression in several passages in the New Testament.[126]

---

[124] S. Safrai, "Teaching of Pietists in Mishnaic Literature," *JJS*, 16 (1965), pp. 15–33; and see Avot de-R. Nathan, Version A, xii, ed. Schechter, p. 56; Midrash ha-Gadol Lev., MS., Shemini, ed. Rabinowitz, p. 240.

[125] Tosefta Yoma i, 12.

[126] Mark 7; Matthew 15; 23; Epistle to the Romans 14:14; etc.

6. Purity and Impurity after the Destruction of the Second Temple

After the destruction of the Second Temple there was a continuing tendency to observe, and to be scrupulous in, matters relating to purity and impurity. This was now done, it appears, with even greater intensity than previously. In and after the generation of Jabneh the Sages adopted the view which extends purity and impurity to include all Israel by not limiting these concepts only to the priests, the Temple, and the holy things.

That the laws of purity and impurity came to occupy such a significant place after the destruction of the Temple, although fundamentally associated with it, may be explained as due to several reasons. i) There was the tendency, based on the belief that the Temple would soon be rebuilt and on Messianic hopes which had then taken root, to refuse to be reconciled to the reality of the destruction of the Temple. ii) In a desire to fill the vacuum created by the destruction of the Temple, the Sages issued many enactments, among them a considerable number relating to purity and impurity. iii) The ascetic tendencies prevalent in Israel consequent on the destruction of the Temple also found expression in a strictness in matters pertaining to purity and impurity. iv) The destruction of the Temple was followed by an increasing study of the Torah, which was prohibited to anyone in a state of ritual impurity.

One of the sources that attest to a stringency in the sphere of purity and impurity after the destruction of the Temple states מיום שחרב בית המקדש נהגו כהנים סילסול בעצמן, שאין מוסרין את הטהרות לכל אדם "From the day the Temple was destroyed the priests guarded their dignity by not entrusting matters of ritual purity to everybody."[127]

Not only the priests but also the laity in Israel were scrupulous in the observance of purity and impurity. A telling description of the tendency to safeguard vigorously the restrictive laws in this area is contained in the tradition in the Jerusalem Talmud which declares תני ר׳ שמעון בן לעזר אומר, ראה עד איכן פרצה טהרה... שלא גזרו לומר שלא יאכל טהור עם הטמא, אלא אמרו לא יאכל הזב עם הזבה "It has been learnt, R. Simeon b. Eleazar said, See how far the observance of ritual purity has spread in Israel... They did not decree that a person who is ritually pure shall not eat with one ritually impure, but they said that a man who has a discharge shall not eat with a woman who has a discharge."[128]

---

[127] TB Bekhorot 30b.

[128] TJ Shabbat i, 3b–c; and parallel passages.

Numerous statements attesting to a strict observance of the laws of purity and impurity by Jews after the destruction of the Temple are to be found in the Church Fathers [see later].

There are evidences that the leading Sages in the period of Jabneh, among them also the Nasi, Rabban Gamaliel of Jabneh, ate secular food in a state of ritual purity. רבן גמליאל היה אוכל על טהרת חולין כל ימיו והיתה מטפחתו מדרס לקודש. אונקלוס הגר היה אוכל על טהרת הקודש כל ימיו והיתה מטפחתו מדרס לחטאת "All his days Rabban Gamaliel ate secular food prepared in ritual purity, and his apron was a source of midras-impurity to those who ate holy things. All his days Onḳelos the proselyte ate secular food prepared according to the degree of purity required for sacred things, and his apron was a source of midras-impurity to those who attended to the water of purification."[129]

There are laws of purity and impurity which by their very nature pertained to the Temple. Nevertheless they continued to be observed. According to one evidence, the priests kept on purifying lepers even after the destruction of the Temple, although it was then quite impossible to offer a sacrifice.[130] And for many generations they continued to purify those impure with corpse-impurity by means of water of purification, using the ashes of the red heifer left over from the days of the Second Temple.[131]

The increase in the batei midrash and the spread of the study of the Torah from the period of Jabneh onwards also brought about the need to observe the laws of purity, since impure persons were prohibited from studying the Torah.[132] An illustration of this is to be found in a source which dates from the days of Jabneh[133] and which tells the following incident about R. Joshua and his disciples תנו רבנן פעם אחת הוצרך דבר

---

[129] Tosefta Ḥagigah iii, 2–3 (and cf. Seder Eliyahu Rabbah xvi, ed. Friedmann, p. 72). And see Lieberman, *Tosefta ki-Peshuṭah*, Seder Mo'ed, pp. 1309–1311.

[130] Tosefta Nega'im viii, 2; Sifra, Metzora', i, 13, ed. Weiss, 70b; TJ Soṭah ii, 18a. There are also evidences showing that they kept aloof from a leper: Lev. Rabbah xvi, 3, ed. Margulies, pp. 352–353. As against these sources, which attest to incidents taken from actual life, a later Midrash contains the statement: א״ר יוחנן משחרב בית המקדש אין טומאת מצורע "R. Johanan said, Since the destruction of the Temple, there is no impurity of the leper" (Leḳaḥ Ṭov, Shemot, vi). This statement was undoubtedly made from the theoretical standpoint.

[131] Tosefta Parah v, 6; etc. And see S. Safrai, "Bet She'arim in Talmudic Literature" (Hebrew), *Eretz-Israel*, 5 (1958), pp. 206–207.

[132] See Allon, *Meḥḳarim*, I, pp. 152–154.

[133] Before the destruction of the Temple there would presumably not have occurred an incident relating to "R. Joshua and his disciples."

אחד לתלמידי חכמים אצל מטרוניתא אחת ... אמרו מי ילך, אמר להם ר׳ יהושע אני אלך... אחר שיצא, ירד וטבל ושנה לתלמידיו, ואמר להם... בשעה שירדתי וטבלתי במה חשדתוני, אמרנו שמא ניתזה צינורא מפיה על בגדיו של רבי, אמר להם העבודה כך היה "Our Rabbis taught, The Sages were once in need of something from a noblewoman... They said, 'Who will go?' 'I will go,' replied R. Joshua... After he came out, he went down, had a ritual bath, and taught his disciples. He said to them... 'When I went down and had a ritual bath, of what did you suspect me?' 'We thought that perhaps some spittle spurted from her mouth upon the Rabbi's garments.' 'By the Temple Service!' he exclaimed to them, 'it was even so'."[134]

In the days of the Severans and during the transition to the subsequent period of anarchy several measures were enacted by the Nesi'im on various subjects with the aim of liberalizing and normalizing the life of the nation religiously, socially, and economically. Against this background a series of lenient enactments, relating to the Sabbatical Year, to tithes, and so on, was issued. A similar series of enactments curtailing the laws of purity and impurity does not exist, although it is natural to assume that the restrictive measures in these areas decreased and diminished in this period. In only one subject in the sphere of purity and impurity do we find leniencies, and that is, in the impurity of gentiles, as embodied in the enactment of R. Judah II Nesi'ah, the grandson of R. Judah ha-Nasi, who permitted the oil of gentiles.[135] Behind this enactment apparently lay the desire to facilitate co-existence with gentiles.[136] There was also a widespread tendency to permit the bread of gentiles.[137] It seems that the leniencies relating to the impurity

---

134 TB Shabbat 127b. The impurity contracted by R. Joshua was the impurity of the gentile, similar in degree to the impurity of one with a discharge. On the impurity of the gentile, see A. Büchler, "The Levitical Impurity of the Gentile in Palestine before the Year 70," *JQR* (N.S.), 17 (1926), pp. 1–81; Allon, *Meḥkarim*, I, pp. 121–147.

135 'Avodah Zarah ii, 6; Tosefta 'Avodah Zarah iv, 11; TJ 'Avodah Zarah ii, 41d; TB 'Avodah Zarah 36a.

136 TJ 'Avodah Zarah ii, 41d contains a tradition which gives the reason for the institution of the enactment as having been due to what occurred: שהיו עולין להר המלך ונהרגין עליו "Because [to obtain pure oil which was not that of the gentiles] they used to go up to the mountain of the king [הר המלך = a Judaean mountain region; see S. Klein, *Eretz Yehudah*, Tel Aviv, 1939, pp. 41–42, 239–247] and were killed when doing so." But this is not to be regarded as the only or the whole reason for permitting the oil.

137 TB 'Avodah Zarah 37a. TB 'Avodah Zarah 35b reports a tradition that already R. Judah ha-Nasi intended to permit the bread of gentiles. There are further evi-

of the gentiles were made possible because of a less stringent attitude to purity and impurity in general.

Despite all that has been said, matters pertaining to purity and impurity were still observed at this period, evidences of their observance being extant both at this time — the end of the tannaitic period — and at the beginning of the days of the Amoraim. In one of his tours of the country R. Judah ha-Nasi was incensed with the inhabitants of a certain place who, although desirous of kneading their dough in purity, did so in impure vessels, having made the mistake of assuming, or having been misled into thinking, that the dough had not been rendered susceptible to impurity.[138] At that time the stringent rule of eating secular food in a state of ritual purity was still held up as an example to be followed, people being urged to put the rule into practice. **רבי חייה רובה מפקד לרב אין את יכיל מיכול כל שתא חולין בטהרה אכול, ואם לאו תהא אכיל שבעה יומין מן שתא** "R. Ḥiyya the Elder charged Rav, If you can eat secular food in a state of ritual purity throughout the year, do so, but if not, eat such food seven days a year."[139] In the amoraic period we find that those impure with corpse-impurity were still purified with water of purification.[140]

The observance of the laws of purity and impurity apparently diminished progressively in the days of the Amoraim, undoubtedly due to the years and generations that had passed since the Temple was in existence and to the crystallization of Judaism around the world of learning and the world of the synagogue and of prayer. It was also due to the diminution and final consumption of the ashes of the red heifer, essential for purification from grave impurities. While a small quantity of the ashes scattered on water was sufficient to produce water of purification, the sprinkling of which purified the impure, in the course of time the "supply" that remained after the destruction of the Temple was clearly exhausted, with no possibility of renewing it.

---

dences of the tendency to make closer contacts with gentiles possible; and see TJ 'Avodah Zarah i, 39b; Tosefta Giṭṭin v, 4; TJ Giṭṭin v, 47c.

138 TB Sanhedrin 5b; TJ Shevi'it vi, 36b–c.

139 TJ Shabbat i, 3c.

140 TJ Berakhot vi, 10a [according to the Rome MS.]: רבי חגיי ורבי ירמיה סלקון למי חטאת קפץ רבי חגיי ובירך עליהן "R. Haggai and R. Jeremiah [Amoraim at the end of the third, the beginning of the fourth, century] went up [to the place of] the water of purification. R. Haggai jumped up and made a blessing over it." On the reading of the Rome MS., see Safrai, *Eretz-Israel*, 5 (1958), pp. 206–207, and note 7 *ad loc.*

CHAPTER THREE

THE 'AM HA-ARETZ LE-MITZVOT AND THE 'AM HA-ARETZ LA-TORAH

1. Introductory Remarks

Talmudic sources themselves endeavoured to define the 'am ha-aretz in various passages which we shall discuss in this chapter. As a rule, every Sage who dealt with the subject sought to point to a particular area that was not properly observed by the 'am ha-aretz. So, for example, we read דתניא, איזהו עם הארץ, כל שאינו אוכל חוליו בטהרה, דברי רבי מאיר, וחכמים אומרים, כל שאינו מעשר פירותיו כראוי "It has been taught, Who is an 'am ha-aretz? Anyone who does not eat his secular food in ritual purity. This is the opinion of R. Meir. The Sages however said, Anyone who does not tithe his produce properly."[1]

Statements such as these, which do not define the 'am ha-aretz as entirely non-observant of the commandments, show that the 'am ha-aretz was part of the nation and that he had no wish to exclude himself, by a rejection of the Torah, from the community of Israel. One has the impression that while the 'am ha-aretz desired, as is the obligation of every Jew, to observe the commandments, he did not devote himself fully and completely to their observance. The Talmud also speaks in positive terms of the 'am ha-aretz's observance of the commandments, as for example, אמר ר׳ שמעון שזורי... שכשם שאימת שבת על עם הארץ, כך אימת דימוע על עם הארץ "R. Simeon Shezuri said, ...Even as the 'am ha-aretz is conscientious in observing the laws of the Sabbath, so is he conscientious in observing the law concerning the mixture of terumah and secular food";[2] or אינו לוקח הימינו לח, הא יבש מותר, שעמי הארץ נאמנין על הכשירות, ותני כן נאמן עם הארץ לומר הפירות הללו לא הוכשרו... "He does not take it from him in a moist state. Hence in a dry state it is permitted, for 'ammei ha-aretz are trustworthy in matters pertaining to food rendered liable to impurity. And so too our Rabbis have

[1] TB Berakhot 47b; and, similarly, TB Giṭṭin 61a [the Munich MS. of Berakhot has כל שאינו אוכל חולין בטהרה "Anyone who does not eat secular food in ritual purity"]; Tosefta 'Avodah Zarah iii, 10.

[2] Tosefta Demai v, 2.

taught, The ʻam ha-aretz is trustworthy when he says, These fruits have not been rendered liable to impurity…";[3] or רוב עמי הארץ מעשרין הן "Most ʻammei ha-aretz separate tithes";[4] and so on. It is of course uncertain to what extent these statements reflect the historical situation in the days of the Second Temple and in the period from its destruction till the first generations of the Eretz Israel Amoraim, and although they often seem to stem from the theoretical discussion in which they occur, it is nevertheless clear that they would not have been made had there not been an explicit tradition that the ʻammei ha-aretz were part of the nation, that they observed some of the commandments, and that while there was a doubt whether they kept others, they were not without the observance of commandments.

The commandments which the ʻam ha-aretz was commonly suspected of not observing related to those associated with the produce of the land in Eretz Israel in general and to tithes in particular, as well as to ritual purity. This is not to suggest that the ʻam ha-aretz scrupulously observed all the other commandments. The choice of these commandments as a means of defining the ʻam ha-aretz by his failure to observe them strictly, and their occurrence in halakhic discussions dealing with the ʻam ha-aretz, were undoubtedly due to three reasons.

a) The centrality of these commandments and the importance attached to their strict observance in the period of the Second Temple and, to a great extent, also after its destruction.

b) Concurrently with the centrality and importance of these commandments their halakhic standing was tenuous — that of the tithes, because of the changes that occurred in the ways of separating them during the days of the Second Temple; and that of ritual purity, because it was not clear whether all Israel were obliged to observe it [see above, pp. 52–55].

c) The non-observance of the commandments relating to tithes and purity could have led others to transgress, whereas if the ʻam ha-aretz did not keep other commandments such as, for example, that of the tzitzit,[5] he personally committed a transgression but this did not lead others coming into contact with him to transgress.

Analogous to the ʻam ha-aretz le-mitzvot is the ʻam ha-aretz la-Torah

---

3 TJ Demai ii, 22d.

4 In the name of Rava in TB Shabbat 23a; in the name of Abbaye in TB Giṭṭin 61a. Cf. also לא נחשדו עמי הארץ על מעשר עני "The ʻammei ha-aretz are not suspected of withholding the tithe of the poor" (TB Nedarim 84b; TB Makkot 17a).

5 TB Berakhot 47b; TB Soṭah 22a.

mentioned in definitions, statements, and discussions in talmudic literature, that is, the 'am ha-aretz in the sense of an ignoramus who has neither studied the Torah nor attached himself to the circles of the Sages.

It may be assumed in principle that more or less the same social stratum comprised alike the 'ammei ha-aretz le-mitzvot and the 'ammei ha-aretz la-Torah, the change from the former to the latter concept having been rather the result of circumstances prevailing at the time and the identity of the particular Sage who defined or discussed the 'am ha-aretz. Consisting of the masses who neither scrupulously observed the commandments relating to tithes and to purity nor engaged in the study of the Torah, this stratum may have incorporated at its fringes those who were 'ammei ha-aretz le-mitzvot but not 'ammei ha-aretz la-Torah, and vice versa.

In general it may be said that the 'ammei ha-aretz la-Torah belonged to the lower classes, to the ordinary people, who did not join in the process which, beginning as early as in the days of the Soferim and continued by the Pharisees, increased in vigour after the destruction of the Second Temple — that process which made the Torah and its study the centre of the world of Judaism. There were naturally also members of the upper classes, "hard men" [takkifim] in cities with mixed populations, as well as others, who failed to become part of this process. Nevertheless the shafts of the Sages were directed at the masses who lived in their midst and were subject to their influence, whom they sought to have as an audience for their Torah, and who were likely to enlarge their congregation.

2. The 'Am ha-Aretz and the Tithes

One of the characteristics defining the 'am ha-aretz describes him as חשוד על המעשרות "suspected with regard to the tithes," that is, suspected of not separating them as prescribed. דתניא איזהו עם הארץ... וחכמים אומרים, כל שאינו מעשר פירותיו כראוי "Our Rabbis taught, Who is an 'am ha-aretz?... The Sages said, Anyone who does not tithe his produce properly."[6] Numerous talmudic sources deal with the different halakhic problems arising from this fact, such as, What is one to do with the produce bought from an 'am ha-aretz? May one

6 TB Berakhot 47b; and, similarly, TB Giṭṭin 61a; Tosefta 'Avodah Zarah iii, 10.

be the guest of an 'am ha-aretz and eat in his home? And so on.

Various factors were responsible for the erosion in the strict observance of the laws pertaining to the tithes.

There was first and foremost the difficulty inherent in the separation of the tithes in that they constituted a heavy tax alongside the other burdensome religious and secular taxes. Nor is there any evidence that the Roman authorities took into consideration the terumot and tithes as they did the Sabbatical Year.[7] The 'am ha-aretz was generally not suspected of neglecting to separate terumah as prescribed, doubtless because of its small amount as compared to the tithes and also because of its firmer halakhic status.[8]

The custom which enjoined that the tithes were to be brought in Temple times to Jerusalem created difficulties for those who lived far away, since, unlike the second tithe, the first tithe could not be redeemed. Generally the tithes were brought to Jerusalem on the occasion of a pilgrimage, but not everyone used to go to Jerusalem on pilgrimage,[9] which was another circumstance that precluded the bringing of the tithes to Jerusalem and even their separation.

The importance that attached to the commandment of the tithes was apparently not apprehended by the 'ammei ha-aretz. For while the priests and the Levites did indeed serve in the Temple, they were far from needing the tithes for their subsistence, as was the case, according to the biblical viewpoint, with the tribe of Levi which had no holding or patrimony. Owning as they did property and estates, and constituting a small percentage of the nation as a whole, the priests and Levites had no need of the tithes of everyone in Israel.

What is more, the Temple service was performed by twenty-four weekly divisions [משמרות] of priests and Levites, each of which thus served for no more than a fortnight in a year, nor presumably did all the members of a division go up to Jerusalem when its turn came to serve in the Temple.

---

7 On taking the Sabbatical Year into consideration, see S. Safrai, "Sabbatical Year Commandments under the Conditions Prevailing after the Destruction of the Second Temple" (Hebrew), *Tarbiz*, 35 (1966), pp. 304–328; 36 (1967), pp. 26–46.

8 So, for example, לפי ששלח בכל גבול ישראל וראה שאין מפרישין אלא תרומה גדולה בלבד, ומעשר ראשון ומעשר שני מקצתן מעשרין ומקצתן אין מעשרין... "Because he [Johanan the high priest] sent [inspectors] throughout Israel and found that they separated only the great terumah, but as for the first and second tithes, some separated them, while others did not..." (TB Soṭah 48a).

9 The subject is dealt with in detail by S. Safrai, *Ha-'Aliyah le-Regel bi-Yemei Bayit Sheni*, Tel Aviv, 1965.

The Pharisaic Sages' dissatisfaction with the way in which the Hasmonaean kings used the tithes undoubtedly contributed to an increase in the number of those who abstained from separating them as prescribed. Not many of the ʻammei ha-aretz followed the bidding of the Pharisaic Sages who directed that the tithes were to be separated and given to a priest [or a Levite] anywhere and were not to be brought to Jerusalem. An expression of the Pharisees' opposition to the Hasmonaean system of collecting the tithes is reflected in the late view that in the days of the Second Temple the obligation relating to the terumot and the tithes was not biblical but מאיליהן קיבלו עליהן "was voluntarily assumed."[10] This view was naturally not conducive to increasing the number of those who separated the tithes, for once an opportunity offered itself of evading the burden of the tithes, even if such was not the intention, many of course took advantage of it.

From the days of Herod onwards the country passed through an extensive process of urbanization. Whereas this affected mainly the non-Jews and the non-Jewish cities, it also had an impact and implications as far as the Jews were concerned, in that the obligation of separating the terumot and the tithes was not imposed upon the townsmen engaged in industry, while on merchants it was binding only in so far as they dealt in agricultural produce from which the sellers had not separated terumot and tithes. With a great deal of probability it may be assumed that the farmers felt that they were burdened with a demand from which the townsmen were exempt, and this feeling of discrimination yet further increased the number of those who evaded the proper separation of the tithes.

Gaining ground progressively up to the end of the Second Temple period, this circumstance of not strictly separating the tithes is clearly reflected in the statement אמרו, חרבו חנויות בני חנן שלש שנים קודם לארץ ישראל, שהיו מוציאים פירותיהם מידי מעשרות, שהיו דורשים לומר, "עשר תעשר ואכלת" — ולא מוכר, "תבואת זרעך" — ולא לוקח "They said, The shops of Benei Ḥanan were destroyed three years before Eretz Israel because they exempted their produce from tithes by interpreting 'You shall tithe... You shall eat the tithe' as implying, but not if you sell it, 'the yield of your seed,' but not if you buy it."[11]

---

10 TJ Sheviʻit v, 36b; and cf. Genesis Rabbah xcvi, ed. Theodor-Albeck, p. 1202; etc.

11 Sifrei Deuteronomy cv, ed. Finkelstein, p. 165; TJ Pe'ah i, 16c; TB Bava Metziʻa 88a. According to E. E. Urbach, "Social and Religious Trends in the Sages' Doctrine of Charity" (Hebrew), *Zion*, 16 (1951), p. 11, note 68, Benei Ḥanan may be dienti-

A Baraita, which includes an appraisal of the Second Temple period, does stress that they were then strict about separating the tithes. אמר רבי יוחנן בן תורתא, מפני מה חרבה שילה, מפני בזיון קדשים שבתוכה, ירושלם בניין הראשון מפני מה חרבה, מפני עבודה זרה וגלוי עריות ושפיכות דמים שהיה בתוכה, אבל באחרונה מכירין אנו בהן שהן עמלין בתורה וזהירין במעשרות, מפני מה גלו, מפני שאוהבין את הממון ושונאין איש את רעהו "R. Johanan b. Torta said, Why was Shiloh destroyed? Because of the disrespect of sacred things there. Why was the First Temple in Jerusalem destroyed? Because of the idolatry, the incest, and the bloodshed that prevailed there. However at the time of the later Temple we know that the people engaged in the study of the Torah and were scrupulous in separating the tithes. Why, then, were they exiled? Because they loved money and hated one another."[12] Most probably the statement of the Baraita that they were זהירין במעשרות "scrupulous in separating the tithes" does not really reflect the actual situation and is rather, so it seems, the consequence of comparing the position in the days of R. Johanan b. Torta [of the third generation of Tannaim] with that in the Second Temple period, when the separation of tithes was nevertheless more widespread than subsequently.

The destruction of the Second Temple naturally played a very significant role in the diminished separation of the tithes both because the service of the priests and the Levites in the Temple had come to an end and because it was no longer possible to bring the tithes to Jerusalem and to the Temple in accordance with the ancient halakhah.

---

fied with Bet Ḥanin, of whom Abba Saul said אוי לי מבית חנין, אוי לי מלחישתן "Woe is me because of Bet Ḥanin, woe is me because of their whisperings" (TB Pesaḥim 57a; and similarly Tosefta Menaḥot xiii, 21). This view is supported by the Berlin MS. of the Sifrei which has the reading חנין, while on the other hand there is also the reading חנן in TB Pesaḥim; see Diḳduḳei Soferim. On the shops of Benei Ḥanan, see Safrai, *Ha-ʿAliyah le-Regel*, pp. 169–170, note 194. A. Büchler, *Der galiläische ʿAm-haʾAreṣ des zweiten Jahrhunderts*, Vienna, 1906, p. 18, note *ad loc.*, advances as proof of the disregard of the separation of tithes at the end of the Second Temple period also the statement in Maʿaserot ii, 5 אמר רבי יהודה, מעשה בגנת ורדים שהיתה בירושלם, והיו תאנים נמכרות משלש ומארבע באסר, ולא הופרש ממנה תרומה ומעשר מעולם "R. Judah said, It happened in a rose garden in Jerusalem that three or four figs were sold for an isar, and neither terumah nor tithe was ever separated from them." But this Mishnah can hardly be used to prove the point, seeing that it occurs in an halakhic context in which R. Judah, expressing the view that figs may be eaten one by one without tithing them, apparently quoted what happened in the rose garden as a precedent from the days of the Second Temple that confirmed his view.

12 Tosefta Menaḥot xiii, 22; and, similarly, in TJ Yoma i, 38c.

[Nevertheless, as pointed out above, pp. 43–49, the tithes were not abolished after the destruction of the Second Temple.] Furthermore, the changes in the ownership of land as a result of the policy adopted by the Roman authorities after the destruction of the Second Temple also presumably led to a diminution in the number of those separating the tithes. Some of the land in Eretz Israel was confiscated and given to ḥokherim,[a] the conductores, or metziḳim[b] in the language of the Jewish sources. These ḥokherim handed over the cultivation of the land to arisim[c] who were in many instances the former owners of the land.[13] Whether under these circumstances the aris had, according to the halakhah, to separate tithes is a question into which we do not now wish to enter, but such confiscations of land clearly led to a decrease in the separation of tithes.

Produce about which there is a doubt whether tithes have been separated from it is called demai, a term that referred specifically to the produce of the ʻam ha-aretz, suspected as he was of not separating the tithes. The linguistic origin of the term is not clear.[14]

The earliest known mention of the word demai occurs in the ordinances of Johanan the high priest: ובימיו אין אדם צריך לשאול על הדמאי

---

[13] On this, see Josephus, *BJ*, vii, 6, 6, §217; Sifrei Deuteronomy ccclvii, ed. Finkelstein, p. 425, and parallel passages; cccxvii, ed. Finkelstein, pp. 359–360; Bava Ḳamma x, 5; TJ Shabbat xvi, 15d; Lamentations Rabbah v, 5. And see G. Allon, *Toledot ha-Yehudim be-Eretz Yisra'el bi-Teḳufat ha-Mishnah ve-ha-Talmud*, Tel Aviv, I: 1958[3], II: 1961[2] — see I, pp. 36–38; Safrai, *Tarbiz*, 35 (1966), pp. 307–310.

[14] As early as in the days of the Amoraim the attempt was made to find the meaning linguistically of the term demai: דמאי, ר׳ יוסי בשם רבי אבהו, רבי חזקיה בשם רבי יודה בר פזי, דמאי – דמי תקן, דמאי לא תיקן "Demai — R. Jose in the name of R. Abbahu, R. Hezekiah in the name of R. Judah bar Pazzi, demai — there is talk that he has given the tithes, there is talk that he has not" (TJ Maʻaser Sheni v, 56d; TJ Soṭah ix, 24b). This explanation sounds rather like a popular etymology. For other suggestions, see *Kaftor va-Peraḥ*, ii, 4a; *Aruch Completum*, s.v. דמאי, and *Additamenta ad Librum Aruch Completum*; Ch. Tchernowitz, "Demai — Hatzaʻah Hisṭorit le-Taḳḳanot Yoḥanan Kohen Gadol she-ba-Mishnah," *Jewish Studies in Memory of G. A. Kohut*, New York, 1935, p. 46, note 1; Lieberman, *Tosefta ki-Peshuṭah*, Seder Zeraʻim, I, p. 192; J. Montgomery, "The Etymology of דמאי," *JQR* (N.S.), 23 (1932–1933), p. 209.

a) Tenant farmers who paid a fixed rent in kind.

b) Roman tax collectors.

c) Tenant farmers who paid a fixed percentage of the crops in rent.

"And in his days one had no need to inquire concerning demai."[15]

In some of its comments on the ordinances of Johanan the high priest, the Talmud associates the term demai with the ʻam ha-aretz's produce that was suspected of not having been tithed. So, for example, **לפי ששלח בכל גבול ישראל, וראה שאין מפרישין אלא תרומה גדולה בלבד, ומעשר ראשון ומעשר שני מקצתן מעשרין ומקצתן אין מעשרין, אמר להם, בני בואו ואומר** לכם כשם שתרומה גדולה יש בה עון מיתה, כך תרומת מעשר וטבל יש בהן עון מיתה. **עמד והתקין להם, הלוקח פירות מעם הארץ מפריש מהן מעשר ראשון ומעשר** שני, מעשר ראשון מפריש ממנה תרומת מעשר ונותנה לכהן, ומעשר שני עולה ואוכלו בירושלים, מעשר ראשון ומעשר עני המוציא מחבירו עליו הראיה. תרתי תקן, ביטל וידוי דחבירים, וגזר על דמאי של עמי הארץ "Because he sent [inspectors] throughout the land of Israel and found that they separated only the great terumah, but as for the first and second tithes some fulfilled the law while others did not, he said to [the people], 'My sons, come, I shall tell you this. Just as in the neglect of the great terumah there is mortal sin, so is there mortal sin in the neglect to present the terumah of the tithe and in the use of ṭevel.'[a] Accordingly he arose and decreed for them that whoever purchases produce from an ʻam ha-aretz must separate from it the first and second tithes. From the first tithe he separates the terumah of the tithe and gives it to a priest, and as for the second tithe he goes up and eats it in Jerusalem. With regard to the first tithe and the tithe of the poor, whoever demands them from his neighbour has the onus of proving [that they had not already been apportioned]. [Johanan] issued two ordinances: he abolished the confession [that the tithes had been duly given] in the case of the ḥaverim and made a decree about the demai of the ʻammei ha-aretz."[16] According to this viewpoint, in Johanan the high priest's ordinance there was a warning to anyone buying produce from an ʻam ha-aretz to separate

---

15 Maʻaser Sheni v, 15; Soṭah ix, 10. Parallel passages have the readings וביטל את הדמאי "And he abolished the demai" (Tosefta Soṭah xiii, 10); גזר על הדמאי "He issued a decree concerning demai" (TB Soṭah 48a). This latter version led some to explain the ordinance as indicating that it was Johanan the high priest [John Hyrcanus] who introduced the concept of demai, and obliged the purchaser to separate from it the tithes. This suggestion is ruled out by the context. It is therefore probable that the concept of demai existed before the days of John Hyrcanus, seeing that he instituted an ordinance which signified a change in respect of demai. The concept must therefore have been in existence some time before the ordinance was promulgated [see also above, pp. 34–35].

16 TB Soṭah 48a.

a) Produce at the stage at which tithes and terumot should be, but have not yet been, separated.

from it, in addition to the terumah, also the terumah of the tithe and the second tithe, but as regards the first tithe and the poor man's tithe, while they have to be designated by name and their location has to be specified, there is no need to separate them in actual fact, since the onus of proof is on the claimant.[17]

From the sources it does not necessarily follow that the concept of the 'am ha-aretz le-ma'aserot existed as early as in the days of the Hasmonaeans, since the viewpoint of the Talmud mentioned above does not reflect the historical conditions at the time the enactment was promulgated. We have previously dealt with the probable significance of the ordinance in the prevailing historical circumstances, with the identity of the legislator, and with its context in the Mishnah, and it has been suggested that in view of the new arrangements for collecting the tithes introduced by John Hyrcanus, arrangements which sought to prevent any possible evasion of separating the tax, there were no longer any misgivings about demai. This ties up with the tradition in the Jerusalem Talmud that שהעמיד זוגות "he [John Hyrcanus] instituted pairs"[18] — that is, state-appointed tithe collectors who worked in pairs.

In view of what has been said it is clear that the term demai may not have been associated originally with the term 'am ha-aretz, but may have referred at first to produce in general about which there was a doubt whether the tithes had been separated, and its association with the 'am ha-aretz suspected of not separating the tithes may have come later. Moreover the actual term demai in the Mishnah in the context of Johanan the high priest's ordinances may even be an anachronism. We are accordingly unable to determine with any certainty whether the concept of the 'am ha-aretz le-ma'aserot existed already in the days of the Hasmonaeans. On the other hand it may be said that the Baraita in tractate Soṭah reflects, as regards the ordinances in the Mishnah, a tradition of interpretation and a viewpoint which presumably did not come into being only after the destruction of the Second Temple. Hence, even if it is not to be assigned as early as to the Hasmonaean period, the concept of the 'am ha-aretz le-ma'aserot is also not to be referred to as late as after the destruction of the Second Temple [as will also

---

[17] In the parallel passages it is not the purchaser who is meant; the ordinance was instituted for the owners and sellers of the produce and was intended to warn them that they were to separate the terumot and tithes in the manner mentioned (Tosefta Soṭah xiii, 10; TJ Ma'aser Sheni v, 56d; TJ Soṭah ix, 24a; in these however the term 'am ha-aretz is not mentioned).

[18] TJ Ma'aser Sheni v, 56d [see above, pp. 34–35].

become apparent from other evidences]. It may be conjectured that the concepts of the ‘am ha-aretz le-ma‘aserot and demai are early, although how early is difficult to determine. They may even have been associated from the outset, their halakhic and social severity increasing with the passage of time.

Because of the laxity of the ‘ammei ha-aretz in separating tithes, a number of halakhot were issued which limited contact with them for fear of transgressing with ṭevel. These limitations refer to two areas: i) dealing in agricultural produce liable to tithes, and ii) being a guest in the home of an ‘am ha-aretz.

Evidence of the extent to which no confidence was reposed in the ‘am ha-aretz in respect of the tithes can be seen from the halakhah which laid down that produce bought from an ‘am ha-aretz had to be tithed, even if he claimed that he had separated the tithes as prescribed. This follows from the Mishnah which declares הלוקח פרות ממי שאינו נאמן על המעשרות, ושכח לעשרן, ושואלו בשבת, יאכל על פיו. חשכה מוצאי שבת לא יאכל עד שיעשר "If a man bought produce from one who is not a ne'eman in respect of tithes, and he forgot to tithe it, he may eat of it at the seller's word if he asked him on the Sabbath. But at the nightfall of the Sabbath day he may not eat of it unless he has first tithed it."[19]

מי שאינו נאמן על המעשרות "One who is not a ne'eman in respect of tithes" is apparently to be identified with an ‘am ha-aretz [for the term ne'eman, see below, pp. 151–156].

The limitations on trading with the ‘am ha-aretz applied also if he was the purchaser. Prohibitions against the sale of produce to the ‘am ha-aretz derive, for example, from the halakhah which states לא ימכר אדם את פרותיו משבאו לעונת המעשרות למי שאינו נאמן על המעשרות "After the season for tithing has arrived, no person may sell his produce to one who is not a ne'eman in respect of tithes."[20]

While the prohibition against being a guest in the home of an ‘am ha-aretz stemmed also, and perhaps principally, from the fear of impurity, the prohibition, it is stressed, was also due to apprehensions about ṭevel. ואל תהי רגיל אצל עם הארץ שסופך להאכילך טבלים "And do not be in the habit of keeping company with an ‘am ha-aretz, for he will eventually give you ṭevalim to eat."[21]

---

19 Demai iv, 1; and also Tosefta Demai v, 1.

20 Ma‘aserot v, 3.

21 TB Nedarim 20a.

Sometimes the members of a family observed the laws relating to the tithes in various degrees of strictness. Where the husband was strict about the tithes and the wife was not, trading with him in produce liable to tithes was permitted, but being a guest in his home was prohibited. הוא נאמן ואשתו אינה נאמנת, לוקחין הימנו ואין מתארחין אצלו, אע״פ שאמרו הרי הוא כשרוי עם נחש בכפיפה אחת "Where he is trustworthy and his wife is not, one may buy [produce] from him but one may not be his guest, although they said that he was like one living in a cage with a serpent."[22] Because the Sages did not approve of such a family, this is one of the instances in which a wife may be divorced without receiving her ketubah.[a)] [23] In the reverse case in which the wife was particular about tithes and the husband was not, it was prohibited to trade with him but permitted to be a guest in his home. אשתו נאמנת והוא אינו נאמן, מתארחין אצלו, ואין לוקחין הימנו "Where his wife is trustworthy and he is not, one may be a guest in his home, but may not buy [produce] from him,"[24] and the Jerusalem Talmud adds אבל אמרו תבוא מאירה למי שאשתו נאמנת והוא אינו נאמן "but they said, May a curse come upon him whose wife is trustworthy and he is not."[25] There are likewise halakhot which refer to families in which the master of the house was not particular about tithes, whereas his son or his slave was, and so on.[26]

An amoraic statement which recurs several times declares that רוב עמי הארץ מעשרין הן "most ʿammei ha-aretz separate the tithes."[27] This statement does not however reflect the situation that obtained in the days of the Tannaim. While the neglect on the part of the ʿammei ha-aretz to separate tithes was due not to an opposition to tithes in principle but to the reasons we have mentioned above, nevertheless most of the ʿammei ha-aretz presumably neglected to separate the tithes, seeing that the very definition of one who belonged to the ʿammei ha-aretz was based, among other factors, on his disregard of the commandments of the tithes. Accordingly this amoraic statement evolved in the course of the discussions in which it occurs and has no reference to the historical

---

22 Tosefta Demai iii, 9; TJ Demai ii, 22d; and see TB Ketubbot 72a.

23 Ketubbot vii, 10.

24 Tosefta Demai iii, 9.

25 TJ Demai ii, 22d.

26 For example, Tosefta Demai iii, 9.

27 In the name of Rava in TB Shabbat 13a and parallel passages; in the name of Abbaye in TB Giṭṭin 61a; etc.

a) Marriage contract, which contains the wife's marriage settlement due to her on being divorced or on her husband's death.

situation. More significant and earlier is the remark in the Tosefta in the name of Rabban Simeon b. Gamaliel הילכות הקודש, תרומה ומעשרות, הן הן גופי תורה ונמסרו לעמי הארץ "The halakhot of heḳdesh, terumot, and tithes are essential parts of the Torah, and were entrusted to the ʿammei ha-aretz."[28] This does not mean that the ʿammei ha-aretz kept these commandments, but rather that the Sages regarded them as central precepts that were to be observed by the ʿammei ha-aretz. In this remark there is also an element, so it seems, of propaganda, the purpose of which was to urge the ʿammei ha-aretz to be scrupulous in observing "the halakhot of heḳdesh, terumot, and tithes," subjects which they were in the habit of disregarding.[29]

In the tannaitic halakhah there are evidences that the ʿammei ha-aretz were suspected of not separating the second tithe, even as they were suspected of not properly observing the separation of the first tithe.[30] On the other hand there are no evidences about the attitude of the ʿam ha-aretz to the poor man's tithe. In the amoraic halakhah there is the statement that לא נחשדו עמי הארץ על מעשר עני "the ʿammei ha-aretz are not suspected of withholding the poor man's tithe."[31] Here too this statement does not necessarily reflect the historical situation, but was apparently made during a theoretical talmudic discussion of the relevant topics. We have thus no actual evidence of the attitude of the ʿammei ha-aretz to the poor man's tithe, although it was presumably not different from their attitude to the first and second tithes. Nor can it be assumed that in the third and sixth years of the Sabbatical cycle they scrupulously separated the poor man's tithe, while neglecting to separate the tithes during the other years.

The halakhah prohibits the giving of terumot and tithes to a priest who is an ʿam ha-aretz. Thus, for example, in the context of laws pertaining to a ḥaver there occurs the injunction ואינו נותן תרומה ומעשר לכהן עם הארץ "And he does not give the terumah and the tithe to a priest who is an ʿam ha-aretz."[32] There is no tendency here to make it obligatory to give the priestly gifts only to a priestly ḥaver in order to pro-

---

[28] Tosefta Shabbat ii, 10; and, similarly, TJ Shabbat ii, 5b; TB Shabbat 32a–b.

[29] The earlier authorities [ha-ri'shonim], seeking to solve the problem of how it could be suggested here that the tithes were sold to the ʿammei ha-aretz when it was known that they were untrustworthy about them, tried to explain it in various ways (Lieberman, *Tosefta ki-Peshuṭah*, Seder Moʿed, pp. 35–36).

[30] In the traditions in the Talmuds on Johanan the high priest's ordinance; etc.

[31] TB Nedarim 84b; TB Makkot 17a.

[32] Avot de-R. Nathan, Version A, xli, ed. Schechter, p. 132.

mote mutual assistance in the association of ḥaverim and to keep the terumot and the tithes to some extent among themselves. The intention is simply to prohibit their being given to a priestly 'am ha-aretz. This we infer from the fact that the halakhah is not worded positively, namely, that he gives the terumah and the tithe to a priestly ḥaver, but negatively in that it prohibits their being given to a priestly 'am ha-aretz. Hence it is permissible to give terumot and tithes to any priest who is not an 'am ha-aretz and who is not necessarily a ḥaver. The reason for the prohibition against giving priestly gifts to a priest who is an 'am ha-aretz is the fear that he might eat them when not in a state of purity. Those scrupulous about purity were particularly strict about observing the purity of terumah. They even ascribed to the observance of the purity of the tithes a greater importance than to the eating of secular food in a state of purity. There is the statement of Amoraim אמר רב אחא בר אדא אמר רב יהודה, כל הנותן תרומה לכהן עם הארץ כאילו נותנה לפני ארי, מה ארי ספק דורס ואוכל, ספק אינו דורס ואוכל, אף כהן עם הארץ ספק אוכלה בטהרה, ספק אוכלה בטומאה "R. Aḥa b. Adda said in the name of R. Judah, Whoever gives terumah to a priest who is an 'am ha-aretz is as though he has placed it before a lion. Even as a lion may possibly tear his prey and eat it and possibly not, so a priestly 'am ha-aretz may possibly eat it in purity and possibly in impurity."[33]

At the end of the tannaitic period there was an insistence on giving the priestly gifts to a "priestly ḥaver," but what is meant here is a ḥaver in the sense of a talmid ḥakham. This marks a transitional stage between giving the terumot and tithes to priests and Levites and giving them to talmidei ḥakhamim who occupied themselves in the study of the Torah.[34]

There was yet another reason for the prohibition against giving terumah to a priest who was an 'am ha-aretz, namely, the fear that he might give it to a non-priest to eat, and for a non-priest the eating of terumah was prohibited. Hence it is stated אל תהי רגיל אצל כהן עם הארץ שסופך להאכילך תרומה "Do not habitually associate with a priest who is an 'am ha-aretz, for he will ultimately give you terumah to eat."[35]

---

33 TB Sanhedrin 90b. Cf. R. Meir's remark about המשיא בתו לעם הארץ "Whoever marries his daughter to an 'am ha-aretz" in TB Pesaḥim 49b.

34 See above, pp. 45–46, 51.

35 TB Nedarim 20a; and cf. Derekh Eretz Rabbah: אבא חלפי אומר משום אבא חגרת אביו... ואל תתארח אצל כהן עם הארץ, שמא יאכילך בקדשי שמים "Abba Ḥilfai said in the name of his father Abba Ḥiggeret... And do not be a guest in the home of a priest who is an 'am ha-aretz, lest he gives you to eat of the sacred things of Heaven"

3. The 'Am ha-Aretz and the Sabbatical Year

The sources in talmudic literature which define the 'am ha-aretz do not identify him as one who disregards or does not observe the commandment of the Sabbatical Year.[36] On the other hand, in a number of sources the tithes and the Sabbatical Year are mentioned together in association with the 'am ha-aretz, such as המוליך חטים לטוחן כותי או לטוחן עם הארץ, בחזקתן למעשרות ולשביעית "If a man took his wheat to a miller who was a Cuthaean or to a miller who was an 'am ha-aretz [the wheat when ground continues] in its former state in respect of tithes and in respect of the laws of the Sabbatical Year."[37] There are moreover several sources in which the expressions חשוד על המעשרות "suspected of not observing the laws relating to the tithes" and חשוד על השביעית "suspected of not observing the laws relating to the Sabbatical Year" occur together. In the previous section [p. 69] we stated that one "suspected of not observing the laws relating to the tithes" and the 'am ha-aretz are to a great extent identical, and this applies also to one "suspected of not observing the laws relating to the Sabbatical Year." As an example of this kind of source there is the statement that לא ימכור אדם את פרותיו משבאו לעונת המעשרות, למי שאינו נאמן על המעשרות; ולא בשביעית, למי שהוא חשוד על השביעית "After the season for tithing has arrived, no person may sell his produce to one who is not a ne'eman in respect of tithes, nor in the Sabbatical Year [may one sell Sabbatical Year produce] to anyone suspected of infringing the Sabbatical Year."[38]

From all this it is evident that the 'am ha-aretz was suspected of not observing the laws of the Sabbatical Year. That this is not mentioned in the sources which define the 'am ha-aretz may be explained by the fact that these sources identify the 'ammei ha-aretz in contrast with the ḥaverim, and since the ḥaver was distinguished by his scrupulous observance of the laws pertaining to the tithes and to purity, the

---

(Derekh Eretz Rabbah i, Tosefta Derekh Eretz iii, 13, ed. Higger, p. 275; Massekhtot Ze'irot, Massekhet 'Arayot xxi, ed. Higger, p. 94).

[36] Demai ii, 3; Tosefta Demai ii, 2–4; TJ Demai iii, 23d; TB Bekhorot 30b; Tosefta 'Avodah Zarah iii, 9; TB Giṭṭin 61a; TB Berakhot 47b; TB Soṭah 22a; TJ 'Eruvin ix, 25d.

[37] Demai iii, 4; TB Giṭṭin 61b. See also Tosefta Demai iv, 22, 24; etc.

[38] Ma'aserot v, 3; and cf. Shevi'it v, 9 (= Giṭṭin v, 9) in which both a woman suspected of not observing the laws of the Sabbatical Year and the wife of an 'am ha-aretz are mentioned.

ʻam ha-aretz's disregard of these laws was by contrast emphasized.[39]

According to Büchler,[40] a greater strictness applied to the ʻam ha-aretz in relation to the Sabbatical Year than in relation to the tithes. For this view of his Büchler bases himself on several sources, such as הוא אינו נאמן, אחד מבניו נאמן, אחד מעבדיו ומשפחותיו נאמן, לוקחין ואוכלין על פיהן, ועושין לו ואוכל — בשביעית ובטהרות אינו רשאי לעשות כן "If [the ʻam ha-aretz] is not trustworthy but one of his sons is trustworthy, or one of his male or female slaves is trustworthy, one may buy and eat [the produce] at their word, and prepare and eat it. But with regard to Sabbatical Year produce and to purity, one is not permitted to do so."[41] There are however sources from which the opposite of Büchler's conclusion is to be deduced: הנותן לחמותו, מעשר את שהוא נותן לה ואת שהוא נוטל ממנה, מפני שהיא חשודה לחלף את המתקלקל. אמר רבי יהודה, רוצה היא בתקנת בתה ובושה מחתנה. מודה רבי יהודה בנותן לחמותו שביעית, שאינה חשודה להאכיל את בתה שביעית "If a man gave produce to his mother-in-law [to prepare as food], he must tithe what he gives to her and what he takes from her, since she is suspected of changing [food] that has been spoiled. R. Judah said, [She may do it because] she is anxious for the well-being of her daughter and is ashamed for her son-in-law. R. Judah agrees that if a man gave to his mother-in-law Sabbatical Year produce, she is not suspected of giving her daughter to eat Sabbatical Year produce."[42]

There is, it seems, no justification for maintaining that a greater strictness was adopted in the case of Sabbatical Year produce than in that of the tithes, seeing that some sources point to a greater strictness in the former, others in the latter, case. These differences were apparently due to divergent attitudes to the Sabbatical Year at various periods, and sometimes in various places during the same period. Such divergent attitudes to the Sabbatical Year may have come from the Sages who for various reasons tended to adopt a lenient or a strict view of the laws of the Sabbatical Year, even as they may have come from the ʻammei ha-aretz whose attitude to the Sabbatical Year and

---

39 Not so Safrai, *Tarbiz*, 35 (1966), p. 325, who contends, in connection with the halakhot which mention one suspected of not observing the laws of the Sabbatical Year alongside one suspected of not separating the tithes and similar things, that there may be no identity between the "suspected," who are mentioned alongside each other only because the same halakhot applied to both of them.

40 *Der galiläische ʻAm-ha-Areṣ*, pp. 213 ff.

41 Tosefta Demai iii, 9; and, similarly, v, 5.

42 Demai iii, 6; TB Giṭṭin 61b; TB Ḥullin 6a.

the tithes was not, in the observance or disregard of them, the same at all periods and in all places. Thus in connection with the last source cited, there is indeed a tradition in the Talmud which declares that באתריה דרבי יהודה חמירא להו שביעית "in the place of R. Judah the Sabbatical Year was strictly observed by the people."[43]

In Mishnah Bekhorot there is evidence that it was possible for a person to be strict about the Sabbatical Year and yet disregard the tithes, and vice versa. החשוד על השביעית אינו חשוד על המעשרות. החשוד על המעשרות אינו חשוד על השביעית. החשוד על זה ועל זה חשוד על הטהרות. ויש שהוא חשוד על הטהרות, ואינו חשוד לא על זה ולא על זה "One who is suspected of disregarding the Sabbatical Year is not suspected of disregarding the tithes. One suspected of disregarding the tithes is not suspected of disregarding the Sabbatical Year. One who is suspected of both is also suspected of disregarding the rules of purity. It is possible for one to be suspected of disregarding the rules of purity and yet not be suspected of disregarding the Sabbatical Year and the tithes."[44]

There are instances in which similar halakhot apply to both the 'am ha-aretz and the tithes, and the 'am ha-aretz and the Sabbatical Year. So, for example, in certain instances it is permitted to deposit produce with an 'am ha-aretz without any misgivings about tithes, even as the sources give cases in which it is permitted to deposit Sabbatical Year produce with an 'am ha-aretz.[45]

To sum up: the conclusion to be drawn from the sources as a whole is that the ordinary Jew was suspected of not observing the laws of the Sabbatical Year.[46] It follows, then, that this applied also to the 'am ha-aretz, as can be learnt directly from the sources too, an example of which is quoted above. The suspicion attaching to the 'am ha-aretz with regard to the Sabbatical Year varied in extent, but was generally greater than was the case with regard to the tithes. Sometimes it was just the opposite. Much depended on the period, on the place, and even on the author of the statement or the halakhah.

---

43 TB Giṭṭin 54a; TB Bekhorot 30a.

44 Bekhorot iv, 10; and cf. Tosefta Bekhorot iii, 12.

45 Tosefta 'Eruvin vii, 10. And cf. Tosefta Demai iv, 26; Demai iii, 4; Tosefta Demai iv, 31; TJ Demai iii, 22d; TB Ḥullin 6a; etc. See also Safrai, *Tarbiz*, 35 (1966), p. 326.

46 See also Tosefta Demai v, 5; TJ Demai iv, 24b; TB Sukkah 39a, with which cf. Tosefta Shevi'it vi, 20; TJ Sukkah iii, 54a; etc.

4. The 'Am ha-Aretz and Ritual Purity

Another area which defines the 'am ha-aretz le-mitzvot — and it is *the* essential and central area in defining him — is his disregard of the laws and restrictions relating to purity. דתניא, איזהו עם הארץ, כל שאינו אוכל חוליו בטהרה, דברי ר׳ מאיר "It has been taught, Who is an 'am ha-aretz? Anyone who does not eat his secular food in ritual purity. This is the opinion of R. Meir."[47] It has been pointed out [see above, p. 17] that in the days of the Second Temple and in the first generations after its destruction there was an increasing concern with ritual purity and impurity as these came to occupy a progressively central place in the Jewish religious consciousness.[48]

A scrupulous regard for matters relating to purity led to a social differentiation based on the various degrees with which the laws and restrictions of purity were observed. The lowest position in the social scale was occupied by the 'ammei ha-aretz, with whom any contact likely to cause impurity was to be avoided.

A social distinction between an element which was faithful and observant of the commandments and one which was unfaithful and lax in its observance of the commandments occurs in the biblical books dating from the early days of the Second Temple period. Mentioned in connection with the making of the firm covenant are "the rest of the people, the priests, the Levites, the gatekeepers, the singers, the temple servants, and all who have separated themselves from the 'ammei ha-aratzot [the peoples of the lands] to the law of God, their wives, their sons, their daughters, all who have knowledge and understanding."[49] The term 'am ha-aretz, signifying here one who was unfaithful, was still far from having the meaning which it acquired towards the end of the Second Temple period.[50] In a description of Passover in the

---

[47] TB Berakhot 47b: the Munich MS. has כל שאינו אוכל חולין בטהרה "Anyone who does not eat secular food in ritual purity"; TB Giṭṭin 61a; Tosefta 'Avodah Zarah iii, 10.

[48] A scrupulous observance of the laws of ritual purity was also prevalent in the pagan world in that period.

[49] Neh. 10:29; cf. Ezra 9:1.

[50] The view current among scholars is that at the beginning of the Second Temple period the term 'am ha-aretz denoted the local population that had remained in the country and had not gone into exile. Between this ancient Israelite population and the returning exiles there was a profound religious contrast. As against the latter, the 'am ha-aretz represented an element that was racially mixed and that had religiously a syncretistic faith. This produced a conflict between the exiles who, purged

Book of Ezra we read: "On the fourteenth day of the first month the returned exiles kept the Passover. For the priests and the Levites had purified themselves together; all of them were clean. So they killed the passover lamb for all the returned exiles, for their fellow priests, and for themselves; it was eaten by the people of Israel who had returned from exile, and also by every one who had joined them and separated himself from the pollutions of the peoples of the land [גויי הארץ] to worship the Lord, the God of Israel."[51] Thus we see that separation from impurity was one of the criteria for determining the social elements that participated in the Passover.

As early as at the end of the third century BCE there were stringent laws relating to ritual purity and impurity, as is evident from the edict of Antiochus III. σεμνύνων δὲ καὶ τὸ ἱερὸν πρόγραμμα κατὰ πᾶσαν τὴν βασιλείαν ἐξέθηκεν περιέχον τάδε· „μηδενὶ ἐξεῖναι ἀλλοφύλῳ

---

in the furnace of repentance, had abandoned idolatry, and the ʻam ha-aretz who had not undergone this process.

Against this view, Y. Kaufmann, *Toledot ha-Emunah ha-Yisra'elit*, Jerusalem & Tel Aviv, 1936–1956, IV, 1, pp. 183–184, contends that there is no difference in the sources between the exiles who returned and the people who remained in the country. According to him, the term ʻam ha-aretz in Haggai (2:4) and in Zechariah (7:5) refers to all the people of Judah, while in Ezra and Nehemiah it denotes foreigners, being analogous to the term "the peoples of the land." On the basis of various verses (Ezra 9:1; Neh. 10:29; and others) I. Seeligmann, *The Septuagint Version of Isaiah*, Leyden, 1948, p. 84, maintains that at the beginning of the Second Temple period as well as in the Books of Maccabees the expression ʻam ha-aretz refers to the peoples living in the border regions of the country which had been conquered by the Jews who compelled them to adopt Judaism. Hated by the Jews, the members of these peoples were called by them by the disparaging name of ʻam ha-aretz. Kaufmann, it should be pointed out, cannot deny that the expression גולה ["the exiles, the returned exiles"] is used in the biblical books dating from the beginning of the Second Temple period to designate a community of faithful Jews (see, for example, Ezra 4:1). Although he contends that this expression is not used in contrast to the people who remained behind in the country, he yet admits that it came into being because the returning exiles were the active force in the nation and the basis of its rebirth and revival, and that it also included those inhabitants of the country who had separated themselves from the peoples and attached themselves to the community of the returning exiles.

There is accordingly no essential difference between Kaufmann's view and that of the other scholars, since he too admits that there were Jews who had separated themselves from the ʻammei ha-aratzot, even as there were others who had not done so. Thus as early as in the days of Ezra and Nehemiah a distinction was made in society between elements who were faithful and those who were close to the foreigners. With these latter there was associated the concept of ʻam ha-aretz.

[51] Ezra 6:19–21.

εἰς τὸν περίβολον εἰσιέναι τοῦ ἱεροῦ τὸν ἀπηγορευμένον τοῖς Ἰουδαίοις, εἰ μὴ οἷς ἁγνισθεῖσίν ἐστιν ἔθιμον κατὰ τὸν πάτριον νόμον..." "And out of reverence for the Temple he also published a proclamation throughout the entire kingdom, of which the contents were as follows. 'It is unlawful for any foreigner to enter the enclosure of the Temple which is forbidden to the Jews, except to those of them who are accustomed to enter after purifying themselves in accordance with the law of the country...'."[52] The one, definite condition mentioned here for entry into the Temple, which was known also to the Seleucid king, was purity.

The Letter of Aristeas, which dates apparently from the end of the second century BCE,[53] in a description of Jerusalem tells that — εἰσὶ δὲ καὶ διαβάθραι πρὸς τὰς διόδους· οἱ μὲν γὰρ μετέωροι τὴν ὁδείαν, οἱ δ' ὑπ' αὐτὰς ποιοῦνται, καὶ μάλιστα διεστηκότας τὰς ὁδείας, διὰ τοὺς ἐν ταῖς ἁγνείαις ὄντας, ὅπως μηδενὸς θιγγάνωσιν ὧν οὐ δέον ἐστίν. "There are steps too which lead up to the cross roads, and some people are always going up, and others down and they keep as far apart from each other as possible on the road because of those who are bound by the rules of purity, lest they should touch anything which is unlawful."[54] Here, then, it is mentioned that in Jerusalem near the Temple there were cross roads, the upper one of which was used by those who were scrupulous about their purity, and the lower one by those who were not, and were accordingly suspected of impurity. Here for the first time there is a clear distinction, which undoubtedly also had a social significance, between those who were particular about purity and impurity and those who were not. For us, those going on the lower cross road are doubtless to be identified with the 'ammei ha-aretz, even though the concept 'am ha-aretz itself did not as yet bear this meaning.

The reference in the Letter of Aristeas to the cross roads in Jerusalem is reminiscent of an halakhic tradition which likewise mentions that there were places in Jerusalem where because of purity and impurity a distinction was made between ḥaverim and 'ammei ha-aretz. This tradition is quoted in the name of Abba Saul who, although active after the destruction of the Second Temple, transmitted many histori-

---

52 *AJ*, xii, 3, 4, §145.

53 E. Bickermann, "Zur Datierung des Pseudo-Aristeas," *ZNW*, 29 (1930), pp. 280–298.

54 Letter of Aristeas, §106.

cally authentic traditions relating to the days of the Second Temple.[55] **אבא שאול אומר, שתי בצעין היה בירושלם התחתונה והעליונה... התחתונה שהיתה קדושתה גמורה עמי הארץ אוכלין בה קדשים קלים** [כ״י וינה ודפוסים: ״ומעשר שני״], **וחברים קדשים קלים אבל לא מעשר שיני, והעליונה שלא היתה קדושתה גמורה עמי הארץ אוכלין בה קדשים קלים אבל לא מעשר שיני, וחבירים לא קדשים קלים ולא מעשר שיני** "Abba Saul said, There were two fissures in Jerusalem, the lower and the upper[56]... In the lower one, which was properly consecrated, 'ammei ha-aretz ate sacrifices of a minor grade of holiness [the Vienna MS. and printed editions: 'and the second tithe'], while ḥaverim ate sacrifices of a minor degree of holiness there but not the second tithe. In the upper one, which was not properly consecrated, 'ammei ha-aretz ate sacrifices of a minor grade of holiness but not the second tithe, whereas ḥaverim ate there neither sacrifices of a minor grade of holiness nor the second tithe."[57] On account of the large number of pilgrims during the days of the Second Temple it was permitted to eat some of the sacred things also outside the Temple precincts, at least in the old parts of Jerusalem.[58] This tradition given in the name of Abba Saul shows that the ḥaverim were scrupulous in eating sacrifices of a minor degree of holiness only in places whose sanctity was certain, and the second tithe only within the Temple precincts, while the 'ammei ha-aretz ate sacrifices of a minor degree of holiness also in places that were not properly consecrated. According to the Erfurt MS. and its parallel versions, the 'ammei ha-aretz ate the second tithe in places whose sanctity was sure; according to the Vienna MS. and the printed editions, they were scrupulous, like the ḥaverim, in eating it only within the Temple precincts. If we adopt the latter version, it should be pointed out that it was possible for the 'ammei ha-aretz to eat the second tithe within the Temple precincts since they were accustomed to bring it to Jerusalem when they made the pil-

---

55 On Abba Saul and his Mishnah, see Israel Lewy, *Über einige Fragmente aus der Mischna des Abba Saul*, Berlin, 1876 [translated into Hebrew in *Mesillot le-Torat ha-Tanna'im*, Tel Aviv, 1928, pp. 92–133].

56 The Vienna MS. has **שתי ביצועין**, from which it appears that the reference is to two fissures and that the word is not derived from **ביצה**, a pond. The parallel passages have **שני ביצעין היו בהר המשחה** "There were two fissures on the Mount of Olives." And see Safrai, *Ha-'Aliyah le-Regel*, p. 153, and notes 234–235.

57 Tosefta Sanhedrin iii, 4; TJ Sanhedrin i, 19b; TB Shevu'ot 16a; Baraita to Megillat Ta'anit vi.

58 See Safrai, *op. cit., loc. cit.*

grimage on the three annual pilgrim festivals and when the impurity of the 'ammei ha-aretz was annulled.[59]

Even as in the non-talmudic halakhic sources, so too in talmudic literature impurity was from ancient times associated with the 'am ha-aretz. In Mishnah Ḥagigah, in a collection of halakhot whose final redaction took place in the days of the Second Temple, the 'am ha-aretz occupied in the scale of the observance of the various degrees of purity the lowest rank.[60] בגדי עם הארץ מדרס לפרושין, בגדי פרושין מדרס לאוכלי תרומה, בגדי אוכלי תרומה מדרס לקדש, בגדי קדש מדרס לחטאת. יוסף בן יועזר היה חסיד שבכהנה והיתה מטפחתו מדרס לקדש, יוחנן בן גדגדא היה אוכל על טהרת הקדש כל ימיו והיתה מטפחתו מדרס לחטאת "The garments of an 'am ha-aretz are a source of midras-impurity to Pharisees, the garments of Pharisees are a source of midras-impurity to those who eat terumah, the garments of those who eat terumah are a source of midras-impurity to those who eat holy things, the garments of those who eat holy things are a source of midras-impurity to those who attend to the water of purification. Jose b. Joezer was the most pious of the priests and his apron was a source of midras-impurity to those who ate holy things. All his life Johanan b. Gudgada ate secular food at the degree of ritual purity required for holy things, and his apron was a source of midras-impurity to those who attended to the water of purification." Thus we see that the social-halakhic concept of the 'am ha-aretz le-ṭohorah crystallized in the days of the Second Temple. This concept is undoubtedly to be identified, in its halakhic, social incidence, with the concept of the 'am ha-aretz le-ma'aserot and la-shevi'it which, we concluded, came into being in the Second Temple period.

In the Mishnah mentioned above, which sets out the degree of ritual impurity attributed to the 'am ha-aretz, the term "midras" is a terminus technicus denoting the pressure of one with a discharge, and is applied to an object on which one with a discharge treads, sits, lies, or leans. Should any of these things be done to the object, it becomes "a father of impurity" [אב הטומאה]. The Mishnah thus compares the impurity of the 'am ha-aretz to that of one with a discharge, and the garments of the 'am ha-aretz are accordingly regarded as "a father of impurity" like the midras of one with a discharge.

Another early Mishnah deals with the impurity of the 'am ha-aretz. כלי חרס מציל על הכל, כדברי בית הלל. ובית שמאי אומרים, אינו מציל אלא על

[59] See below, pp. 93–95.

[60] Ḥagigah ii, 7; and see above, pp. 59–60.

האכלין, ועל המשקין, ועל כלי חרס. אמרו להם בית הלל, מפני מה, אמרו להם בית שמאי, מפני שהוא טמא על גב עם הארץ, ואין כלי טמא חוצץ, אמרו להם בית הלל, והלא טהרתם אכלים ומשקין שבתוכו, אמרו להם בית שמאי, כשטהרנו אכלים ומשקין שבתוכו, לעצמו טהרנו, אבל כשטהרת את הכלי, טהרת לך ולו, חזרו בית הלל להורות כדברי בית שמאי "An earthenware vessel can, according to the opinion of Bet Hillel, protect everything [in it from contracting impurity engendered by a corpse that is under the same roof]. Bet Shammai however said, It protects only foodstuffs and liquids and [other] earthenware vessels. Bet Hillel asked them, Why? Bet Shammai replied, Because it is impure on account of an ‘am ha-aretz. And no impure vessel can screen [against impurity]. Bet Hillel replied, And did you not pronounce pure the foodstuffs and liquids inside it? Bet Shammai answered, When we pronounced pure the foodstuffs and liquids inside it, we pronounced them pure [only] for [the ‘am ha-aretz] himself, but when you pronounced the vessel pure you pronounced it pure for yourself and for him. Thereupon Bet Hillel retracted and decided in accordance with the opinion of Bet Shammai."[61] This Mishnah forms part of the collection and redaction of the halakhah undertaken in the generation of R. Johanan b. Zakkai, following the destruction of the Second Temple, with the object of reducing divergences of opinion so as to restore as far as possible unity to the Torah of the Sages.[62] The share of this generation in the redaction of the Mishnah consisted mainly in coordinating the halakhot in which there had originally been a controversy between Bet Hillel and Bet Shammai, but in which Bet Hillel later, still in the days of the Second Temple, changed their view and decided like Bet Shammai. Of this circumstance, the present Mishnah is an example. To determine these issues R. Johanan b. Zakkai's generation was assisted by Sages who were active in the days of the Second Temple and were thus able to testify concerning an issue such as the one under discussion. Since this Mishnah mentions the ‘am ha-aretz in connection with purity and impurity, and since it clearly dates from the Second Temple period and not just from its last days seeing that it refers to two developments in the days of the Second Temple, this once more indicates the early date of the concept of the ‘am ha-aretz le-ṭohorah and its use in the historical, halakhic, and social situation during the period of the Second Temple. From this Mishnah it also follows that the ‘am ha-aretz was not entirely without keeping things associated with purity

[61] ‘Eduyyot i, 14; TB Ḥagigah 22a–b. And see A. Büchler, "Die Schammaiten und die levitische Reinheit des עם הארץ," *Freimann-Festschrift*, Berlin, 1937, pp. 21–37.

[62] See Tosefta ‘Eduyyot i, 1.

and impurity but that he had his own defined concepts which differed from those of Sages, Pharisees, and ḥaverim, being, as they were, below the norm of purity demanded by them.

The use of the concept ʿam ha-aretz in matters relating to purity did not come to an end with the destruction of the Second Temple but continued during the generations of the Tannaim. Even in the early days of the Eretz Israel Amoraim we find it used. Thus it is told that Resh Laḳish, coming to the place of de-Bei R. Yannai,[63] said to them עם הארץ אני אצל הטהרות "I am an ʿam ha-aretz as regards ritual purities."[64] Hence the concept was in active use as late as the middle of the third century CE. From this source we may indeed also infer the tenuousness of the concept at that time, seeing that an important Amora such as Resh Laḳish included himself among the ʿammei ha-aretz as regards ritual purity. So tenuous had the subject of purity and impurity become at the beginning of the amoraic period that it was limited only to individuals and to special groups, such as that of de-Bei R. Yannai.[65] [66]

A large number of halakhot concern the ritual impurity of the ʿam ha-aretz, its degree, the apprehensions entertained of it, and so on. The ʿam ha-aretz was regarded as impure in the degree of "a father of impurity," that is to say, like one who had been rendered impure by a corpse and had taken no steps to purify himself. His impurity was likened to that of one with a discharge, while to his wife there was attributed the impurity of a menstruant. Accordingly ʿammei ha-aretz conveyed impurity by contact and by carriage, by shaking [hesseṭ] and by pressure [midras].

We came across this definition of an ʿam ha-aretz's impurity in part for the first time in the Mishnah in Ḥagigah: בגדי עם הארץ מדרס לפרושין

63 An association which, living at ʿAkhbarah, scrupulously observed purity.

64 TJ Demai ii, 23a.

65 See E. E. Urbach, *Ḥazal — Pirḳei Emunot ve-Deʿot*, Jerusalem, 1969, p. 571, and note 23 *ad loc.*

66 Indeed already in the days of the Second Temple there was a Sage upon whom no reliance was placed in matters pertaining to ritual purity and impurity. מעשה ברבן גמליאל הזקן שהשיא את בתו לשמעון בן נתנאל הכהן ופסק עמו על מנת שלא תעשה טהרות על גביו "It was told of Rabban Gamaliel the Elder that he married his daughter to Simeon b. Nethanel the priest and stipulated with him that it was on condition that she was not to prepare food for him in ritual purity" (Tosefta ʿAvodah Zarah iii, 10). If the reference is to Simeon b. Nethanel, the pupil of R. Johanan b. Zakkai (Avot iii, 8), we have here a case of a talmid ḥakham about whom doubts were had in connection with matters relating to purity and impurity.

"The garments of an ʿam ha-aretz are a source of midras-impurity to Pharisees."[67] It is also inferred from many halakhic sources that deal with the impurity of the ʿam ha-aretz, with the care to be exercised against it, and with the relationship between the ʿam ha-aretz and the ḥaver.[68] For example, המפקיד כלים אצל עם הארץ טמאים טמא מת וטמאין מדרס "If a man deposited articles with an ʿam ha-aretz they are deemed to be impure with corpse-impurity and with midras-impurity."[69] This halakhah is the consequence of comparing the impurity of the ʿam ha-aretz with that of one impure with corpse-defilement and of one with a discharge. The articles deposited with the ʿam ha-aretz become impure at the degree of corpse-impurity on the principle that articles which touch a person impure with corpse-impurity become, like him, "a father of impurity."[70]

Another halakhah which defines the ʿam ha-aretz both as impure with corpse-defilement and as one who has a discharge states חבר ועם הארץ שהיו שרויין בחצר ושכחו כלים בחצר, הראוי ליטמא מדרס מטמא מדרס, וליטמא טמא מת, טמא טמא מת "A ḥaver and an ʿam ha-aretz who live in the same courtyard and forgot articles in the courtyard, whatever is liable to become midras-impure causes midras-impurity, to become corpse-impure causes corpse-impurity."[71]

The impurity of the ʿam ha-aretz and its degree were established and were not the subject of dispute among the Sages. On the other hand numerous halakhot deal with the problem of the extent to which misgivings were to be entertained about the impurity of the ʿam ha-aretz. For example, המניח עם הארץ בתוך ביתו, ער ומצאו ער, ישן ומצאו ישן, ער ומצאו ישן, הבית טהור. ישן ומצאו ער הבית טמא, דברי רבי מאיר. וחכמים אומרים, אין טמא אלא עד מקום שהוא יכול לפשוט את ידו ולגע "If a man left an ʿam ha-aretz in his house awake and found him awake, or asleep and found him asleep, or awake and found him asleep, the house remains pure. If he left him asleep and found him awake the house is impure. This is the opinion of R. Meir. But the Sages said, Only that part is impure which, by stretching out his hand, he can touch."[72] From this halakhah a central principle is to be derived, the importance of which extends

---

[67] From this Mishnah it follows that the ritual impurity of the ʿam ha-aretz involves not only his body but also his garments. Cf. Tosefta Demai ii, 15.

[68] See below, pp. 161–169.

[69] Ṭohorot viii, 2.

[70] See Oholot i, 3.

[71] Tosefta Ṭohorot ix, 1.

[72] Ṭohorot vii, 2.

beyond an halakhic analysis of the 'am ha-aretz's status, the principle that the 'am ha-aretz is not suspected of wilfully making things impure, for under ordinary circumstances in which the householder left him in the house, he is not suspected, even according to R. Meir whose view is the stricter one, of touching articles and making them impure. There are a number of other halakhot in the same spirit, such as המוסר מפתחו לעם הארץ, הבית טהור, שלא מסר לו אלא שמירת המפתח "If a man entrusted his key to an 'am ha-aretz, the house remains pure, since he entrusted him only with guarding the key."[73] However in the case in which the master of the house left the 'am ha-aretz asleep and found him awake, and there is the fear that when he awoke he moved around in the house, a difference of opinion exists, as we have seen, between R. Meir and the Sages on the extent to which fears are to be entertained that he has defiled the house, with R. Meir adopting, as usual, the stricter view of the halakhot relating to the 'am ha-aretz and ruling that the house is impure, and with the Sages taking the lenient view and holding that it is impure only as far as he can stretch out his hand and touch things.[74]

The above-mentioned halakhah which starts with המפקיד כלים אצל עם הארץ continues אם מכירו שהוא אוכל בתרומה, טהורין מטמא מת, אבל טמאין מדרס "If he knew that he ate terumah, they [the articles] are free from corpse-impurity but are impure with midras-impurity."[75] The reference here is to an 'am ha-aretz who was acquainted with the depositor and knew him to be a priest who ate terumah and was scrupulous about ritual purity. In such a case it is known that the 'am ha-aretz would be careful not to defile the articles, which are consequently free from corpse-impurity, but since there is the fear that they may have been unintentionally defiled, as, for example, if the 'am ha-aretz's wife sat on them during the period of her menstruation, they are impure. Here we can clearly see that the 'am ha-aretz did not deliberately and spitefully defile things and that in certain instances he was even prepared to be especially careful not to make someone else's articles impure.[76]

---

[73] Ṭohorot vii, 1. However Tosefta Ṭohorot viii, 1 has רבי שמעון אומר, המוסר מפתח לעם הארץ הבית טמא וכו׳ "R. Simeon said, If a man entrusted a key to an 'am ha-aretz, the house is impure..." See also TJ Pesaḥim ii, 28d; TJ Ḳiddushin i, 60b.

[74] Cf. also Ṭohorot vii, 4, 5; Tosefta Ṭohorot v, 7; etc.

[75] Ṭohorot viii, 2.

[76] The Mishnah is quoted in *Teshuvot Ge'onim Ḳadmonim* 20b, where it is explained that the articles are midras-impure מפני שבגדי אוכלי תרומה מדרס לקדש "Since the garments of those who eat terumah are a source of midras-impurity to those

At times the word of the 'am ha-aretz was relied upon in matters associated with ritual purity. והתניא נאמן עם הארץ לומר פירות לא הוכשרו, אבל אינו נאמן לומר פירות הוכשרו אבל לא נטמאו "It has been taught, An 'am ha-aretz is trusted if he says that the produce has not been rendered susceptible to impurity. He is however not trusted if he says that the produce has been rendered susceptible to impurity but has not become impure."[77] So long as the 'am ha-aretz said that the produce in his possession had not been rendered susceptible to impurity, the produce was held to be ritually pure, and there was no fear that his statement was untrue and made to mislead others or to increase his profits. But no reliance was placed upon him if he said that the produce had been rendered susceptible to impurity but had not been made impure, the reason being apparently that he was not suspected of telling an untruth but rather of not taking the necessary care or of not having an adequate knowledge of the subjects of purity and impurity.[78]

Sometimes the 'am ha-aretz was trusted in matters of purity and impurity not because he could be relied upon from the outset in a particular case but to prevent a widening in the nation of the social gap created by the fear of contact with the 'am ha-aretz because of misgivings about his impurity.[79]

Two cases of this kind are particularly notable.

The one case: עם הארץ שאמר טהור אני לחטאת, מקבלין אותו; כלים הללו טהורין לחטאת, מקבלין הימנו "An 'am ha-aretz who said, 'I am ritually pure for a sin-offering,' is accepted; 'these vessels are pure for a sin-offering,' they are accepted from him."[80] The reason for the trustworthiness of the 'am ha-aretz in this instance is undoubtedly to be found in R. Jose's statement in an earlier halakhah. וכן היה רבי יוסי אומר, מפני מה הכל נאמנין על החטאת, ואין הכל נאמנין לא על הקודש ולא על התרומה, שלא יהא כל אחד ואחד אומר הריני בונה מזבח לעצמי, הריני שורף את הפרה לעצמי, שנאמר, "אתה ובניך אתך תשמרו את כהונתכם לכל דבר המזבח וגו'" "And so R. Jose said, Why is everybody trusted about a sin-offering but not

who eat holy things." This is however unacceptable. (*Otzar ha-Ge'onim*, Ḥagigah, p. 41.)

[77] TB Ḥagigah 22b.

[78] Similarly, Tosefta 'Avodah Zarah iv, 11, which however has also this exception אבל אין נאמן לומר דגים אילו צדתים בטהרה, ולא ניערתי עליהן את המכמרות "But he is not trusted if he says, I caught these fish in purity and did not shake the nets over them" (see Makhshirin v, 7).

[79] See the next chapter [pp. 118–169], *passim*, and especially pp. 164 ff.

[80] Tosefta Ḥagigah iii, 22; and cf. Tosefta Parah iv, 12–14.

everybody is trusted about holy things or about terumah? So that everyone should not say, I shall build an altar for myself, I shall burn a red heifer for myself, as it is said, ‘And you and your sons with you shall attend to your priesthood for all that concerns the altar...’.”[81] A unified worship was a sufficient reason to override, in the case mentioned in the halakhah, the fear of the ʿam ha-aretz’s impurity. To the same context belongs apparently the following halakhah in Mishnah Ḥagigah, even though the term ʿam ha-aretz is not mentioned in it. חמר בתרומה, שביהודה נאמנים על טהרת יין ושמן כל ימות השנה, ובשעת הגתות והבדים אף על התרומה וכו׳ “Greater strictness applied to terumah [than to holy things], for in Judaea they were trusted as regards the purity of holy wine and oil throughout the year and as regards terumah only at the season of the wine-presses and the olive-vats.”[82] For this halakhah, too, R. Jose assigned the same reason as he did for trusting the ʿam ha-aretz in the case of the sin-offering. דתניא, אמר רבי יוסי, מפני מה הכל נאמנין על טהרת יין ושמן כל ימות השנה, כדי שלא יהא כל אחד ואחד הולך ובונה במה לעצמו, ושורף פרה אדומה לעצמו “It has been taught, R. Jose said, Why is everyone trusted throughout the year about the purity of the wine and the oil [they bring for Temple use]? So that everyone should not go and build a high place for himself and burn a red heifer for himself.”[83]

The second of these two cases deals with the annulment during a festival of the ʿam ha-aretz’s impurity. טומאת עם הארץ ברגל כטהרה שויוה רבנן “The Rabbis declared the impurity of the ʿam ha-aretz to be purified on a festival.”[84] Here, too, the reason could indeed be that on a festival everyone was more scrupulous about his ritual purity than during the rest of the year.[85] And yet presumably all ʿammei ha-aretz did not purify themselves for a festival. What is more, during a festival it was obligatory to be much more particular than usual in matters connected with purity, since Jerusalem and the Temple were involved. In view of

---

81 Tosefta Ḥagigah iii, 19: in the Vienna MS., the first printing, and the London MS., in the name of R. Nehemiah. According to Lieberman, *Tosefta ki-Peshuṭah*, Seder Moʿed, pp. 1324–1325, the correct version of this halakhah is אמר רבי יוסי מפני מה הכל נאמנין על החטאת ועל הקדש ואין הכל נאמנין על התרומה וכו׳, “R. Jose said, Why is everyone trusted about a sin-offering and about the holy things but everyone is not trusted about terumah...” (Yalḳuṭ Talmud Torah, MS., Num. 18:7 in the name of the Tosefta).

82 Ḥagigah iii, 4.

83 TB Ḥagigah 22a.

84 TB Betzah 11b; TB Ḥagigah 26a.

85 See G. Allon, *Meḥḳarim be-Toledot Yisraʾel*, Tel Aviv, 1957–1958, I, pp. 156–157; Safrai, *Ha-ʿAliyah le-Regel*, pp. 135–141.

all this a greater significance attaches to those halakhot which lay down that during a festival any misgivings about the 'am ha-aretz's impurity fell away in respect of various matters. The aim of these halakhot was undoubtedly the social unity of the nation by preventing any barriers on those days which by their very nature symbolized the social and national unity of the people, thereby enabling everyone to make the pilgrimage and thus participate as much as possible in the experiences associated with it. This atmosphere found expression in the remark of R. Joshua b. Levi.[86] אמר רבי יהושע בן לוי, דאמר קרא "ויאסף כל איש ישראל אל העיר כאיש אחד חברים" – הכתוב עשאן כולן חברים "R. Joshua b. Levi said, The biblical verse declares, 'So all the men of Israel gathered against the city, united [ḥaverim] as one man.' — The text made them all ḥaverim."[87] This means that during a festival the status of the 'ammei ha-aretz was analogous to that of ḥaverim who were scrupulous in observing, and adopted a strict view of, the laws of purity and impurity.[88]

Proof that during a festival the impurity of the 'ammei ha-aretz was annulled even though they were definitely still impure is to be found in the regulations associated with the Temple services following a festival. משעבר הרגל היו מעבירין על טהרת העזרה... כיצד מעבירים על טהרת העזרה, מטבילין את הכלים שהיו במקדש וכו' "As soon as the festival was over they cleared up for the purification of the Temple court... How did they clear up for the purification of the Temple court? They immersed the vessels which were in the Temple..."[89] There was clearly the fear that the 'ammei ha-aretz had during the festival defiled the Temple vessels, and although the 'ammei ha-aretz were regarded as ritually pure during the festival, after it there was the fear that the vessels might have become impure retrospectively. Accordingly they acted to restore the purity of the Temple to its former state.[90]

---

86 Of the first generation of Eretz Israel Amoraim.

87 TB Ḥagigah 26a; and, similarly, TJ Ḥagigah iii, 79d; Midrash Psalms, Ps. 122, ed. Buber, p. 508; TJ Bava Ḳamma vi, 6a; TB Niddah 34a.

88 From this, the halakhah in Tosefta Ḥagigah iii, 34 becomes clear: בירושלם נאמנין על טהרת הכלים לקודש, ובשעת הרגל אף על התרומה "In Jerusalem they are trusted as regards the purity of vessels for holy things, and during a festival also as regards terumah"; and so too Ḥagigah iii, 6, with the latter part of the Mishnah connected with its beginning which deals with trustworthiness concerning the purity of earthenware vessels.

89 Ḥagigah iii, 7–8.

90 The Mishnah apparently refers to an action for the purpose of purification. The expression מעבירים is not clear. Some explain it as מעבירים קול, that is, they

The purity of the 'ammei ha-aretz on a festival held good not only as regards their entering the Temple. It applied also to the entire range of possible contacts with them on a festival. It is against this background that the following halakhah is to be understood. הפותח את חביתו, והמתחיל בעיסתו, על גב הרגל, רבי יהודה אומר, יגמר, וחכמים אומרים, לא יגמר "If a man opened his barrel or broke into his dough on account of the festival, R. Judah says he may finish [selling them after the festival], but the Sages say he may not finish [selling them after the festival]."[91]

Two cases are dealt with here. The one refers to a shopkeeper who was scrupulous in matters relating to purity and impurity and opened a barrel of wine to sell it during the festival. The second case deals with a man, likewise scrupulous as regards purity and impurity, who began to sell dough on the festival. In both cases his customers included 'ammei ha-aretz who on normal days would have made the barrel of wine and the dough impure. There was no argument about the barrel and the dough during the festival itself, for everyone agreed that they were then ritually pure. It was only about the days following the festival that the question arose whether the state of purity passed on to them or not. On this question R. Judah and the Sages disagreed. According to R. Judah's view, the barrel and the dough persisted in their state of purity also after the festival, whereas according to the Sages' outlook the purity did not continue after the festival, as we have seen previously also in the case of the Temple vessels.[92]

In several matters relating to purity and impurity it was permitted to be associated throughout the year with an 'am ha-aretz, the reason for this being the same as in previous cases, namely, the fear of enlarging the barrier between the 'ammei ha-aretz and the other sections of the nation. At times it is stressed that the reason for such permission is מפני דרכי שלום "in the interests of peace." So, for example, משאלת אשה לחברתה החשודה על השביעית: נפה, וכברה, ורחים, ותנור; אבל לא תבר ולא

proclaimed that the Temple court was to be purified, others that it means removing the vessels from their places in order to immerse them and purify the Temple court, others again explain it as meaning that they hurried and endeavoured to arrive early for the work. See Ch. Albeck, Commentary to the Mishnah, "Hashlamot ve-Tosafot," *ad loc.* Perhaps the word is to be connected with העברת הטומאה, "removing the impurity," but then this is difficult in the context of מעבירים על טהרת העזרה.

91 Ḥagigah iii, 7.

92 The term על גב הרגל "on account of the festival" means, then, based on the annulment of the 'am ha-aretz's impurity during the festival.

תטחן עמה. אשת חבר משאלת לאשת עם הארץ: נפה, וכברה; ובוררת, וטוחנת, ומרקדת עמה; אבל משתטיל המים לא תגע עמה, לפי שאין מחזיקין ידי עוברי עברה. וכלן לא אמרו אלא מפני דרכי שלום. ומחזיקין ידי גוים בשביעית — אבל לא ידי ישראל — ושואלין בשלומן, מפני דרכי שלום "A woman may lend to another who is suspected of not observing the Sabbatical Year a fan or a sieve or a handmill or a stove, but she should not sift or grind with her. The wife of a ḥaver may lend to the wife of an ʻam ha-aretz a fan or a sieve and may winnow and grind and sift with her, but once she has poured water over the flour she should not touch anything with her, because it is not right to assist those who commit a transgression. All these rules were laid down only in the interests of peace. Heathens may be assisted in the Sabbatical Year — but not Israelites — and greetings may be given to them in the interests of peace."[93]

Despite all that has been said, the most scrupulous care was taken with regard to the impurity of the ʻam ha-aretz.[94] Of all the apprehensions entertained about the ʻam ha-aretz the most stringent and insistent was the fear of the impurity that attached to the ʻam ha-aretz, to his clothes, and to his things. In certain doubtful instances in which there were no misgivings about the separation of tithes and about the observance of the Sabbatical Year, there yet persisted the fear of the ʻam ha-aretz's impurity. For example, ורמינהו, המוליך חטין לטוחן כותי או לטוחן עם הארץ, הרי אלו בחזקתן למעשר ולשביעית, אבל לא לטומאה "A further contradiction was raised. If a man takes wheat to a miller who is a Cuthaean or to a miller who is an ʻam ha-aretz, it is presumed to remain in its original condition as regards the tithe or Sabbatical Year produce, but not as regards impurity."[95] On the other hand, one who was trustworthy with regard to purities was naturally also trustworthy with regard to the other things which the ʻam ha-aretz was suspected of neglecting. תני, הנאמן על הטהרות, נאמן על המעשרות "It is learnt, Whoever is trustworthy about purities is trustworthy about tithes."[96]

---

93 Giṭṭin v, 9.

94 We shall revert to this when we deal with the status of the ʻam ha-aretz in society and with his attitude to those sections in it which strictly observed purity: see below, pp. 156–169.

95 TB Giṭṭin 61b.

96 TJ Demai ii, 22d.

5. The 'Am ha-Aretz and the Torah

Analogous to the 'am ha-aretz le-mitzvot is the 'am ha-aretz la-Torah, that is to say, the unlearned ignoramus.

The 'am ha-aretz la-Torah and le-mitzvot — that is, the commandments associated with the study of the Torah — occurs in the following Baraita which seeks to define the identity of the 'am ha-aretz. תנו רבנן איזהו עם הארץ, כל שאינו קורא קריאת שמע ערבית ושחרית דברי רבי אליעזר. רבי יהושע אומר, כל שאינו מניח תפילין. בן עזאי אומר, כל שאין לו ציצית בבגדו. רבי נתן אומר, כל שאין מזוזה על פתחו. רבי נתן בן יוסף אומר, כל שיש לו בנים ואינו מגדלם לתלמוד תורה. אחרים אומרים, אפילו קרא ושנה ולא שמש ת״ח הרי זה עם הארץ "Our Rabbis taught, Who is an 'am ha-aretz? Anyone who does not recite the shema' evening and morning. This is the view of R. Eliezer. R. Joshua said, Anyone who does not put on tefillin. Ben 'Azzai said, Anyone who does not have tzitzit on his garment. R. Nathan said, Anyone who does not have a mezuzah at his door. R. Nathan b. Joseph said, Anyone who has sons and does not bring them up to a study of the Torah. Others again said, Anyone who has learnt Scripture and Mishnah but has not ministered to talmidei ḥakhamim is an 'am ha-aretz."[97]

Here we have depicted a different image of the 'am ha-aretz. It is no longer the 'am ha-aretz who is not particular about the commandments of the tithes and of purity, but the 'am ha-aretz who, shunning the study of the Torah and the world of the Sages, is designated as such because of his ignorance. The commandments which are mentioned in the Baraita, and of the non-observance of which various Sages accused the 'am ha-aretz, are all associated directly or indirectly with a reference to the study of the Torah and with an emphasis on the importance of such study. This applies to the recital of the shema', to the tefillin, to the tzitzit, and to the mezuzah. Particularly far-reaching is the conclusion of the Baraita which lays down that even if a man has engaged in the study of the Torah, so long as he has not ministered to talmidei ḥakhamim he is still regarded as an 'am ha-aretz, which means that a man has to be intimately associated not only with the world of the Torah but also with that of the Sages.

It is not surprising that the Sages mentioned in the Baraita are for the most part Tannaim of the generation of Jabneh, the generation in

---

97 TB Berakhot 47b; and cf. TB Soṭah 21b–22a; and see Urbach, *Ḥazal*, p. 571, and the notes *ad loc.*

which, as a result of the destruction of the Second Temple, as a result, too, of the striving to fill the vacuum created by that destruction, the study of the Torah increased enormously in value and acquired a central and supreme significance, similar to the observance of purity and of the tithes in the days of the Second Temple.

In the period of Jabneh following the destruction of the Second Temple there arose the need to consolidate and unite the nation around religious aims and objectives that could be achieved also in days when the Temple no longer existed. Coupled with this was the attempt to continue to maintain as much as possible the commandments and customs that prevailed during the Second Temple period. The factor with whose help the leadership came to unite and consolidate the people, and the objective and aim which were set before every Jew, were found in the world of the Torah and of its study, and in the world, too, of the Sages. From the days of Jabneh onwards the estimate of a man was based on the extent to which he engaged in the study of the Torah, on his knowledge of it, and on his belonging to the circle of the Sages or to the disciples ministering to their teachers. As a result a new social stratum was created with the talmid ḥakham at the top of the social scale and at its bottom the ignoramus who had not studied the Torah and who shunned talmidei ḥakhamim. He was given an opprobrious title, that of 'am ha-aretz.[98]

There are numerous sources dating from the destruction of the Second Temple to the first generations of the Eretz Israel Amoraim which deal with the 'am ha-aretz in this sense of one who had not studied and was not particular about the Torah, and which contrast the 'am ha-aretz with the talmid ḥakham.[99]

R. Eliezer b. Hyrcanus, who was a leading Sage of Jabneh and one of the most distinguished pupils of R. Johanan b. Zakkai in the days of the Second Temple and in the period of Jabneh, described in the following statement the impact that the destruction of the Second Temple had on the nation's social strata according to the criterion of their relation to the Torah. ר׳ אליעזר הגדול אומר, מיום שחרב בית המקדש, שרו חכימיא למהוי כספריא, וספריא כחזניא, וחזניא כעמא דארעא, ועמא דארעא אזלא ונדלדלה, ואין מבקש — על מי יש להשען, על אבינו שבשמים "R. Eliezer the Great said, From the day the Temple was destroyed the Sages began

[98] On the relations between the 'ammei ha-aretz and the talmidei ḥakhamim, see below, pp. 170–199.

[99] On the position during the period of the Second Temple, see below, p. 103.

to be like school teachers, the school teachers like synagogue-attendants, the synagogue-attendants like the ʻam ha-aretz, and the ʻam ha-aretz became more and more debased; there was none to ask. Upon whom are we to rely? Upon our Father who is in heaven."[100] As can be seen from this statement, the Sage was at the top of the social scale, and in contrast to him the ʻam ha-aretz — the reference here is naturally to the ʻam ha-aretz la-Torah — was at the bottom.[101]

Another of R. Eliezer's statements set the Sage in contrast to the ignorant ʻam ha-aretz. תניא, רבי אליעזר אומר, זקנים שבדור כהוגן שאלו, שנאמר, "תנה לנו מלך לשפטנו," אבל עמי הארץ שבהן קלקלו, שנאמר, "והיינו גם אנחנו ככל הגוים ושפטנו מלכנו ויצא לפנינו" "It was taught, R. Eliezer said, The elders of the generation made a fitting request, as it is written, 'Give us a king to govern us,' but the ʻammei ha-aretz among them acted unworthily, as it is written, 'That we also may be like all the nations, and that our king may govern us and go out before us'."[102] It is clear that R. Eliezer b. Hyrcanus here compared the situation in his day with what took place at the establishment of the kingdom in the days of Samuel.[103]

---

[100] Soṭah ix, 15. This Mishnah is one of the Baraitot "additions" to the Mishnah. See J. N. Epstein, *Mavo le-Nusaḥ ha-Mishnah*, Jerusalem, 1964², II, pp. 949, 976–977. Lowe's MS., the Munich MS., and ʻEin Yaʻaḳov, the first printed edition, have it in the name of R. Joshua; Seder Eliyahu Zuṭa, ed. Friedmann, Pirḳei Derekh Eretz, i, p. 8, has it in the name of R. Eliezer b. Jacob, the full version there being: רבי אליעזר בן יעקב אומר, מיום שחרב בית המקדש תבו חכימיא למיהוי כספרייא וספרייא כדורשיא ודרשיא כתלמידים ותלמידים כעמא ארעא ועמי ארעא כעמא ועמא אזלן נזלו "R. Eliezer b. Jacob said, From the day the Temple was destroyed the Sages became like school teachers, the school teachers like lecturers, the lecturers like disciples, the disciples like ʻammei ha-aretz, the ʻammei ha-aretz like the people, and the people degenerated continually." In all its different versions the statement was made by a Tanna of the generation of Jabneh.

[101] It is not quite clear whether the ʻam ha-aretz in its conceptual-social sense is meant here or whether the reference is to the members of the nation in general, seeing that the Mishnah uses the feminine נדלדלה. In Seder ʻOlam Zuṭa the ʻammei ha-aretz are mentioned separately from the nation, from which it is clear that the reference is to the ʻam ha-aretz in the conceptual-social sense, but this version appears to be totally corrupt.

[102] TB Sanhedrin 20b.

[103] The Munich MS. has R. Eliezer b. Zadok; other MSS. and Tosefta Sanhedrin iv, 5 have R. Eliezer b. R. Jose. If the reference is to R. Eleazar b. Zadok the First, whose father, R. Zadok, was one of the heroes of the aggadot on the destruction of the Second Temple, the statement still belongs to the generation of Jabneh, but if the reference is to R. Eleazar b. R. Jose, then it is later and belongs to the days of R. Judah ha-Nasi.

Another statement which dates from the period of Jabneh and which deals with the 'am ha-aretz la-Torah is the remark of R. Dosa b. Harkinas who was active at the end of the days of the Second Temple and in the period of Jabneh. ר׳ דוסא בן הרכינס אומר, שנה של שחרית, ויין של צהרים, ושיחת הילדים, וישיבת בתי כנסיות של עמי הארץ, מוציאין את האדם מן העולם "R. Dosa b. Harkinas said, Morning sleep, midday wine, children's talk, and sitting in the synagogues of the 'ammei ha-aretz put a man out of the world."[104] In this statement there is already an explicit undertone of contempt for the 'ammei ha-aretz.[105]

Statements and halakhot that deal with the 'am ha-aretz la-Torah and that generally contrast him with the talmid ḥakham increased in the generation of Usha and continued in the days of R. Judah ha-Nasi as well as in the first generations of the Eretz Israel Amoraim.

For example, from the generation of Usha: אמר רבי אלעזר, למה תלמיד חכם דומה לפני עם הארץ, בתחלה דומה לקיתון של זהב, סיפר הימנו דומה לקיתון של כסף, נהנה ממנו דומה לקיתון של חרש, כיון שנשבר שוב אין לו תקנה "R. Eleazar[106] said, How is the talmid ḥakham regarded by the 'am ha-aretz? At first, like a golden ladle; if he converses with him, like a silver ladle; if he [the talmid ḥakham] derives benefit from him, like an earthen ladle, which once broken cannot be mended."[107]

An instructive comparison between the aged 'ammei ha-aretz and the aged talmidei ḥakhamim occurs in the following statement. רבי שמעון בן עקשיא אומר, זקני עם הארץ כל זמן שמזקינין דעתן מטרפת עליהן, שנאמר, ״מסיר שפה לנאמנים וטעם זקנים יקח.״ אבל זקני תורה אינן כן, אלא כל זמן שמזקינין דעתן מתישבת עליהן, שנאמר, ״בישישים חכמה וארך ימים תבונה״ "R. Simeon b. 'Aḳashya[108] said, The aged among the 'am ha-aretz, the older they become the more their intellect gets distracted, as it is said, 'He deprives of speech those who are trusted, and takes away the discernment of

---

104 Avot iii, 10. For the alternative reading כנסיות "assemblies," see above, p. 21, note 76.

105 See R. Ishmael b. Eleazar's statement in TB Shabbat 32a. We revert to this on pp. 173–174, below.

106 R. Eleazar by itself refers to R. Eleazar b. Shammua' who was of the generation of Usha and a pupil of R. Akiva. Printed versions have R. Eliezer, but this is unacceptable; see Diḳduḳei Soferim.

107 TB Sanhedrin 52a–b.

108 This is the only extant statement by R. Simeon b. 'Aḳashya, and it is therefore difficult to determine his date. Some maintain that he was the brother of R. Hananiah b. 'Aḳashya who apparently belonged to the generation of Usha. A similar statement was made by R. Ishmael b. Jose, a contemporary of R. Judah ha-Nasi: see TB Shabbat 152a.

the elders,' whereas with aged scholars it is not so; on the contrary, the older they get the more their mind becomes composed, as it is said, 'Wisdom is with the aged, and understanding in length of days'."[109]

R. Meir and R. Judah, who were among the most prominent Sages of Usha, often dealt with the 'am ha-aretz le-ma'aserot who was not particular about the commandments associated with the produce of the land in Eretz Israel and about purity, and contrasted him with the ḥaver who was scrupulous and strict in these matters. These two Sages also dealt with the 'am ha-aretz la-Torah,[110] and as usual compared him with the talmid ḥakham. דרש ר׳ יהודה בר׳ אלעאי, מאי דכתיב, ״שמעו דבר ה׳ החרדים אל דברו״ — אלו תלמידי חכמים, ״אמרו אחיכם״ — אלו בעלי מקרא, ״שנאיכם״ — אלו בעלי משנה, ״מנדיכם״ — אלו עמי הארץ; שמא תאמר פסק סברם ובטל סיכוים, ת״ל ״ונראה בשמחתכם״; שמא תאמר ישראל יבושו, ת״ל ״והם יבושו״ — עובדי כוכבים יבושו, וישראל ישמחו "R. Judah b. Ila'i taught, What is meant by the verse, 'Hear the word of the Lord, you who tremble at his word'? This refers to the talmidei ḥakhamim; 'your brethren... have said,' these are the students of Scripture; 'who hate you,' this refers to students of the Mishnah; 'and cast you out,' these are the 'ammei ha-aretz. Lest you say, Their hope is destroyed and their prospects frustrated, Scripture says, 'That we may see your joy.' Lest you think Israel shall be ashamed, therefore it is said, 'It is they who shall be put to shame': the idolators shall be ashamed, whereas Israel shall rejoice."[111]

R. Meir's acrimonious remark about the man who marries his daughter to an 'am ha-aretz refers to an 'am ha-aretz who is devoid of the Torah. תניא היה ר׳ מאיר אומר כל המשיא בתו לעם הארץ, כאילו כופתה ומניחה לפני ארי, מה ארי דורס ואוכל ואין לו בושת פנים, אף עם הארץ מכה ובועל ואין לו בושת פנים "It has been taught, whoever marries his daughter to an 'am ha-aretz is as though he bound and laid her before a lion; just as a lion tears [his prey] and devours it and has no shame, so an 'am ha-aretz strikes and cohabits and has no shame."[112] In this context there is

---

109 Ḳinnim iii, 6.

110 In a parallel version to the identification of the 'am ha-aretz in Berakhot 47b: תנו רבנן, איזהו עם הארץ, כל שאינו קורא קריאת שמע שחרית וערבית בברכותיה, דברי רבי מאיר "Our Rabbis taught, Who is an 'am ha-aretz? Anyone who does not recite the shema' morning and evening with its blessings. This is the view of R. Meir" (TB Soṭah 22a).

111 TB Bava Metzi'a 33b [here ultimately is a favourable attitude to the 'am ha-aretz: we shall revert to this. He is of course contrasted with the talmid ḥakham].

112 TB Pesaḥim 49b; and cf. TB Sanhedrin 90b.

another remark made by R. Judah. רבי יהודה אומר, עם הארץ הרי הוא כגוי, שנאמר, "את בתך לא תתן לבנו" "R. Judah said, The 'am ha-aretz is like a gentile, as it is said, 'You shall not [give] your daughters to their sons'."[113]

The assigning of the 'am ha-aretz to the lowest rank of the social scale in contrast to the talmid ḥakham who had status and enjoyed numerous privileges continued also in the generation of R. Judah ha-Nasi. So, for example, it is told of R. Judah ha-Nasi himself that רבי פתח אוצרות בשני בצורת, אמר — יכנסו בעלי מקרא, בעלי משנה, בעלי גמרא, בעלי הלכה, בעלי הגדה; אבל עמי הארץ אל יכנסו "Rabbi [R. Judah ha-Nasi] once opened his storehouse [of victuals] in a year of scarcity, proclaiming, Let those enter who have studied Scripture, or Mishnah, or gemara, or the halakhah, or the haggadah, but let not the 'ammei ha-aretz enter."[114]

The succession of statements dealing with the 'am ha-aretz la-Torah continued until the days of the second and third generations of Eretz Israel Amoraim. R. Johanan, the senior Amora in the second generation, said: אמר רבי יוחנן, כל החלומות הולכין אחר פתרוניהן, חוץ מן היין. יש שותה יין וטוב לו, יש שותה יין ורע לו — תלמיד חכם שותה וטוב לו, עם הארץ שותה ורע לו "R. Johanan said, All dreams follow their interpretations, except in the case of wine. One drinks wine and it is well with him; another drinks wine and it is bad with him. A talmid ḥakham drinks wine and it is well with him, an 'am ha-aretz drinks wine and it is bad with him."[115] R. Ḥama b. Ḥanina of the third generation of Eretz Israel Amoraim said: אמר רבי חמא ברבי חנינא, מה תקנתו של מספרי לשון הרע, אם תלמיד חכם הוא יעסוק בתורה, שנאמר, "מרפא לשון עץ חיים," ואין לשון אלא לשון הרע, שנאמר, "חץ שחוט לשונם," ואין עץ אלא תורה, שנאמר, "עץ חיים היא למחזיקים בה," ואם עם הארץ הוא ישפיל דעתו, שנאמר, "וסלף בה שבר ברוח" "R. Ḥama b. Ḥanina said, What is the remedy for slanderers? If he is a talmid ḥakham, let him engage in the Torah, as it is said, 'The healing for a tongue is the tree of life' — and 'tongue' here means the evil tongue, as it is said, 'Their tongue is a deadly arrow,' and 'tree [of life]' means the Torah, as it is said, 'She is a tree of life to those who

113 Pirḳa de-Rabbenu ha-Ḳadosh, Bava de-Arba'ah, ed. Schönblum, 21b. There are corresponding statements that advise a man to marry the daughter of a talmid ḥakham and not of an 'am ha-aretz. See TB Pesaḥim 49a. We revert to the subject on pp. 172–173, below.

114 TB Bava Batra 8a. We shall later deal with the subject in detail.

115 TJ Ma'aser Sheni v, 55c; and cf. Genesis Rabbah lxxxix, 7, ed. Theodor-Albeck, pp. 1096–1097.

lay hold of her.' But if he is an 'am ha-aretz, let him become humble, as it is said, 'But perverseness in it breaks the spirit'."[116]

It is clear that the concept of the 'am ha-aretz la-Torah became filled with content and significance after the destruction of the Second Temple. The question is whether it also came into being then, or whether it existed already in the days of the Second Temple, even though at that time it undoubtedly would have had less force and content than after the destruction of the Temple, when the study of the Torah and its students became so central as to constitute the very heart of the national-religious-social life of the nation.

In general it may be said that the study of the Torah was regarded as a positive value of prime importance likewise in the days of the Second Temple, during all of which there were not only students of the Torah and Sages, but illiteracy and ignorance were undoubtedly considered a defect. This is evident even from non-talmudic sources. Thus, for example, Josephus was proud of the extensive knowledge of the Torah and of the laws possessed by the people. ἡμῶν δ' ὁντινοῦν τις ἔροιτο τοὺς νόμους ῥᾷον ἂν εἴποι πάντας ἢ τοὔνομα τὸ ἑαυτοῦ. τοιγαροῦν ἀπὸ τῆς πρώτης εὐθὺς αἰσθήσεως αὐτοὺς ἐκμανθάνοντες ἔχομεν ἐν ταῖς ψυχαῖς ὥσπερ ἐγκεχαραγμένους, καὶ σπάνιος μὲν ὁ παραβαίνων, ἀδύνατος δ' ἡ τῆς κολάσεως παραίτησις. "But, should anyone of our nation be questioned about the laws, he would repeat them all more readily than his own name. The result, then, of our thorough grounding in the laws from the first dawn of intelligence is that we have them, as it were, engraven on our souls. A transgressor is a rarity; evasion of punishment by excuses an impossibility."[117]

There is extant a source which dates, so it would seem, from the days of the Second Temple, and in which the 'am ha-aretz is apparently mentioned as one who has not studied the Torah. We refer to the saying of Hillel אין בור ירא חטא, ולא עם הארץ חסיד "An uncultured person is not sin-fearing, neither is an 'am ha-aretz pious."[118] This saying is based on a sort of literary parallelism between "an uncultured person" and "an 'am ha-aretz" and between "sin-fearing" and "pious."[119] According to this, there existed as early as in the days of Hillel, under Herod, the concept of the 'am ha-aretz la-Torah.

---

116 TB 'Arakhin 15b.

117 *Against Apion*, ii, 18, §178.

118 Avot ii, 5.

119 See Urbach, *Ḥazal*, pp. 522–523.

But this is not so certain, since it may be that the saying is not to be ascribed to Hillel the Elder. In MSS. of the Mishnah, in Mishnayot from the Genizah, and in a MS. of Maimonides' Mishnah commentary the reading is רבי הלל אומר "R. Hillel said."[120] If we accept this version, the saying is to be ascribed to R. Hillel, the grandson of R. Judah ha-Nasi.[121]

The mention in the Mishnah of R. Hillel who was the grandson of R. Judah ha-Nasi and who belonged to the third century CE is possible, seeing that other contemporary Sages are also quoted in the Mishnah, such as, for example, Ben Hé Hé who is mentioned in tractate Avot itself,[122] or R. Judah [II] Nesi'ah who was likewise a grandson of R. Judah ha-Nasi.[123]

The Sages cited in Avot ii before the sayings given in the name of Hillel or R. Hillel also reinforce the possibility that we have here R. Hillel of the third century CE and not Hillel the Elder. The chapter opens with sayings of R. Judah ha-Nasi, continues with those of Rabban Gamaliel his son, after which come those of Hillel or R. Hillel, the sayings being apparently arranged here in order of lineal descent: R. Judah ha-Nasi, his son R. Gamaliel, his grandson R. Hillel.[124]

Nor are the contents of the saying particularly appropriate to Hillel the Elder. The saying here deals with a sin-fearing and with a pious person. Hillel the Elder himself was not one of the pious men of olden days [חסידים הראשונים] nor one of the pious men and men of action [חסידים ואנשי המעשה], having been rather representative of the type of Sage who was indeed virtuous but not specifically pious in the true sense of the concept.[125]

---

120 So in Lowe's, the Parma, and Kaufmann's MSS., and in the Berlin MS. of Maimonides' Mishnah commentary, as well as in Mishnayot from the Genizah, Antonin Collection (Ginzei Mishnah, pp. 102–103), and in the MS. of Maḥzor Vitry. See also Avot, ed. Taylor.

121 See A. Hyman, *Toledot Tanna'im ve-Amora'im*, London, 1910, s.v. – הלל ב' בן רבן גמליאל בר רבי; Epstein, *Mavo le-Nusaḥ ha-Mishnah*, I, pp. 46–47, II, pp. 1182–1183; S. Safrai, "Teaching of Pietists in Mishnaic Literature," *JJS*, 16 (1965), pp. 15–33.

122 Avot v, 23. It is interesting that the same Ben Hé Hé is mentioned in conversation with Hillel: א"ל בר הי הי להלל "Bar Hé Hé said to Hillel" (TB Ḥagigah 9b). The reference here is undoubtedly to R. Hillel, and Bar Hé Hé is none other than Ben Hé Hé (see Hyman, *op. cit.*, s.v. בר הא הא).

123 'Avodah Zarah ii, 6.

124 The Sages mentioned before Hillel or R. Hillel were Nesi'im, which suggests that the reference may after all be to Hillel the Elder.

125 On the person of Hillel the Elder, see E. E. Urbach, *Ha-Entziḳlopedyah ha-'Ivrit*, XIV, p. 537, s.v. הלל; and also *idem*, *Ḥazal*, pp. 513–530.

Accordingly, we are inclined to conclude that the author of the saying was R. Hillel, from which it follows that the concept of the ‘am ha-aretz la-Torah did not exist in the days of the Second Temple. But even in the less likely possibility that it was uttered by Hillel the Elder, if we go more deeply into the saying it becomes evident that it does not embody the concept of the ‘am ha-aretz la-Torah in the same way in which it occurs after the destruction of the Second Temple. In this saying there is indeed a certain parallelism between an ‘am ha-aretz and an uncultured person, according to which the ‘am ha-aretz is identified with one who has not studied. But what is said about the ‘am ha-aretz is ולא עם הארץ חסיד "neither is the ‘am ha-aretz pious." Thus we see that the ‘am ha-aretz is condemned not because he is not a talmid ḥakham but because he is not pious, that is to say, he is not distinguished for doing good in the life of the community, such as redeeming captives, returning a lost article, giving charity, comforting mourners, and so on.[126] The contrast between the ‘am ha-aretz and the pious man is akin to the contrast between the ‘am ha-aretz and the ḥaver in the days of the Second Temple, since the associations of ḥaverim in the Second Temple period were distinguished not only for the scrupulous observance of the tithes and of purity but also for good deeds within the sphere of communal life, after the manner of the pious [החסידים].[127]

If R. Hillel is the author of the saying, then the idea contained in it is no novelty seeing that it occurs in all the generations in which the ‘am ha-aretz la-Torah was dealt with, from the days of Jabneh until those of the first generations of the Eretz Israel Amoraim.

A similar statement is quoted in the name of R. Akiva. אין עם הארץ חסיד, לא הביישן למד, ולא הקפדן מלמד "An ‘am ha-aretz is not pious, nor does a bashful person learn, nor does an impatient person teach."[128]

A biting statement in this context was made in the generation of Usha in the name of R. Simeon b. Yoḥai. עם הארץ — אפילו חסיד, אפילו ישרן, אפילו קדוש ונאמן, ארור הוא לאלהי ישראל "An ‘am ha-aretz — even

[126] See Safrai, *JJS*, 16 (1965), pp. 15–33.

[127] See below, pp. 140–142.

[128] Avot de-R. Nathan, Version A, xxvi, ed. Schechter, p. 82; and similarly Version B, xxxiii, ed. Schechter, p. 72: אין בור ירא חטא, ולא כל המרבה בסחורה מחכים, ולא עם הארץ חסיד "An uncultured person is not sin-fearing, nor does everyone who engages much in trade become wise, nor is an ‘am ha-aretz pious." Instead of חסיד "pious," the Halberstam MS. has פרוש "a Pharisee."

a pious one, even an upright one, even a holy and trustworthy one, cursed be he unto the God of Israel."[129] Thus even if the 'am ha-aretz is pious and performs good deeds in the life of the community, this still does not remove him from the ranks of the 'ammei ha-aretz, for the intention here is undoubtedly that so long as the 'am ha-aretz does not study the Torah and become a talmid ḥakham, virtues will not help him even if he rises to the level of being pious.

Similar to this is a statement made at the beginning of the amoraic period. ואמר רבי אבא אמר ר׳ שמעון בן לקיש, אם תלמיד חכם נוקם ונוטר כנחש הוא חגריהו על מתניך, אם עם הארץ חסיד, אל תדור בשכונתו "R. Abba said in the name of R. Simeon b. Laḳish, Even if a talmid ḥakham is vengeful and bears malice like a serpent, gird him to your loins, [whereas even] if an 'am ha-aretz is pious, do not dwell in his vicinity."[130]

The words אין עם הארץ חסיד "an 'am ha-aretz is not pious" do not follow a different line from that embodied in the last two statements, since these were uttered ironically. Accordingly it is not to be inferred from them that pious 'ammei ha-aretz increased in number, but rather that they were not pious. For these statements are intended to convey the notion that even should the 'ammei ha-aretz become pious, they cease to be 'ammei ha-aretz only if they also become talmidei ḥakhamim.

6. The Status of the 'Am ha-Aretz la-Torah in the Halakhah

Even as the lack of the scrupulous observance by the 'am ha-aretz of the laws relating to tithes and to purity necessitated the introduction of numerous halakhot in order to prevent his causing others to transgress through ṭevel or through impurity, so was there also created in the halakhah a special status for the 'am ha-aretz la-Torah in order to preclude any possibility of someone else or the community being led to transgress through the 'am ha-aretz's ignorance of the Torah.

A clear distinction between the 'ammei ha-aretz on the one hand and the talmidei ḥakhamim on the other occurs in connection with their appointment as judges in civil cases. The original halakhah laid down that a bet din which dealt with civil cases could be composed of three lay judges. In the period however of Jabneh there developed the ten-

---

129 Pirḳa de-Rabbenu ha-Ḳadosh, Bava de-Arba'ah, ed. Schönblum, 21b.
130 TB Shabbat 63a.

dency to have talmidei ḥakhamim and not 'ammei ha-aretz act as judges also in civil cases, since the very fact of their being talmidei ḥakhamim gave them the status almost of competent judges. It is against this background that we have to view the halakhah "אשר תשים לפניהם": לפני בני ישראל לא נאמר כן, אלא לפנים שבהם, מלמד שאין שונין בדיני ממונות לעם הארץ ["'Now these are the ordinances] which you shall set before them': it does not say here 'before the children of Israel,' but [before] the important persons among them, which means that civil laws are not to be taught to the 'am ha-aretz."[131]

The tension between, on the one hand, the ancient halakhah which placed the dispensing of justice also in the hands of laymen and, on the other, the tendency in the period of Jabneh to assign it only to talmidei ḥakhamim is conspicuous in the following difference of opinion between two Tannaim of the generation of Jabneh. "ושפטו את העם בכל עת" — רבי יהושע אומר, בני אדם שבטילין ממלאכתן יהיו דנין את ישראל בכל עת. רבי אלעזר המודעי אומר, בני אדם שבטילין ממלאכתן ועסיקין בתורה יהו דנין את ישראל בכל עת "'And let them judge the people at all times': R. Joshua said, Persons who abstain from work shall judge Israel at all times. R. Eleazar of Modi'im said, Persons who abstain from work and engage in the study of the Torah shall judge Israel at all times."[132] Whereas R. Joshua's criterion for a person to act as a judge was the economic one, the reason being that it was prohibited for a judge to receive payment for acting in that capacity, R. Eleazar of Modi'im was not satisfied only with qualifications in the economic sphere and demanded of a candidate for the position of a judge that he also engaged in the study of the Torah, that is, that he came from the class of the talmidei ḥakhamim.[133]

The endeavour to place the administration of justice exclusively in the hands of talmidei ḥakhamim was also due in the period of Jabneh to the desire of the leadership to increase its authority throughout the country. The tendency to assign this function solely to talmidei ḥakhamim undoubtedly led to its being given to competent judges, who were ordained as such by the central body at Jabneh.[134] This naturally created

---

131 Mekhilta de-R. Simeon b. Yoḥai on Ex. 21:1, ed. Epstein-Melamed, p. 158.

132 Mekhilta de-R. Simeon b. Yoḥai on Ex. 18:22, ed. Epstein-Melamed, p. 133; and parallel passages.

133 See Allon, *Toledot ha-Yehudim*, I, p. 140.

134 According to the source which deals with the development of ordination, every Sage in the period of Jabneh could, so it would seem, ordain his pupils. אמר רבי

an additional means whereby in effect the central institution at Jabneh ruled over the Jewish population in the country.

An attempt to make a distinction in another halakhic subject between the ʻam ha-aretz and the talmid ḥakham occurred in one of the clashes between R. Joshua b. Hananiah and the Nasi Rabban Gamaliel of Jabneh.[135] R. Joshua wanted to lay down that if the priest owning the firstling was a talmid ḥakham, he himself could decide whether it had a blemish, and if so, declare it permitted for slaughter. רבי צדוק הוה ליה בוכרא, רמא ליה שערי בסלי נצרים של ערבה קלופה, בהדי דקאכיל איבזע שיפתיה. אתא לקמיה דרבי יהושע, אמר ליה, כלום חילקנו בין חבר לעם הארץ, אמר לו רבי יהושע, הן "R. Zadok had a firstling. He set down barley for it in wicker baskets of peeled willow twigs. As it was eating, its lip was slit. He came before R. Joshua and asked him, Have we made any difference between [a priest] who is a ḥaver and [a priest] who is an ʻam ha-aretz? R. Joshua replied, Yes."[136] According to R. Joshua's verdict, a difference was made between a priest who was a ḥaver in the sense of his being a talmid ḥakham and a priest who was an ʻam ha-aretz la-Torah in that the former was authorized to declare firstlings permissible for himself, whereas the latter was not authorized to do so. R. Joshua's verdict was not accepted, and it was repeated that every priest required permission to slaughter a firstling from one who had received the necessary authorization. The rejection of R. Joshua's view was not

---

בא, בראשונה היה כל אחד ואחד ממנה את תלמידיו, כגון רבן יוחנן בן זכיי מינה את ר׳ ליעזר ואת רבי יהושע, ורבי יהושע את רבי עקיבא, ורבי עקיבא את רבי מאיר ואת רבי שמעון וכו׳ "R. Ba [= Abba] said, At first each teacher used to ordain his own pupils. Thus R. Johanan b. Zakkai ordained R. Liezer and R. Joshua, and R. Joshua [ordained] R. Akiva, and R. Akiva [ordained] R. Meir and R. Simeon..." (TJ Sanhedrin i, 19a). The examples however of the Sages who ordained their pupils show clearly that those who were permitted to confer ordination were the most important members of the Sanhedrin. This source emphasizes the Sanhedrin's authority to confer ordination which at this stage was greater than that of the Nasi. In view of this it is possible to understand the continuation of this statement which declares that in the following period it was the Nasi who conferred ordination and which begins with the words חזרו וחלקו כבוד לבית הזה "They once more honoured this house." This source may also reflect the tension that existed between R. Johanan b. Zakkai and his pupils on the one hand and the house of the Nasi of the dynasty of Hillel on the other (and see Allon, *Toledot ha-Yehudim*, I, p. 142; *idem*, *Meḥkarim*, I, pp. 272–273, and notes 85–86 *ad loc.*).

135 On the clashes between Rabban Gamaliel and R. Joshua, see Allon, *Toledot ha-Yehudim*, I, pp. 197–201; L. Ginzberg, *Perushim ve-Ḥiddushim ba-Yerushalmi*, New York, 1941–1961, III, pp. 168–220.

136 TB Bekhorot 36a.

because of any opposition to the discrimination in the halakhah between the 'am ha-aretz and the talmid ḥakham, but because the Nasi wished to retain under his jurisdiction the right to declare firstlings permissible for slaughter. Thus it is stated on this subject that דבר זה הניחו להם לבי נשיאה כדי להתגדר בו "this matter was left in the hands of the Nasi as a special distinction for him."[137]

In areas in which there was a question of honour and of showing respect to others it was emphatically declared that talmidei ḥakhamim always took precedence over 'ammei ha-aretz. כהן קודם ללוי, לוי לישראל, ישראל לממזר, וממזר לנתין, ונתין לגר, וגר לעבד משחרר. אימתי, בזמן שכלן שוין, אבל אם היה ממזר תלמיד חכם וכהן גדול עם הארץ — ממזר תלמיד חכם קודם לכהן גדול עם הארץ "A priest takes precedence over a Levite, a Levite over an Israelite, an Israelite over a mamzer,[a)] a mamzer over a Natin,[b)] a Natin over a proselyte, and a proselyte over an emancipated slave. This order of precedence applies only when all these are in other respects equal. But if a mamzer is a talmid ḥakham and the high priest an 'am ha-aretz, the mamzer who is a talmid ḥakham takes precedence over the high priest who is an 'am ha-aretz."[138] This view was held as early as in the days of R. Johanan, who was of the second generation of Eretz Israel Amoraim, when in Sepphoris there arose the problem who was to take precedence in greeting the Nasi. תרין זרעיין הוון בציפרין בולווטיא ופגניא, הוון שאלין בשלמיה דנשייא בכל יום, והוון בולווטיא עלין קדמיי ונפקין קדמיי, אזלון פגניא וזכון באוריתא אתון בעו מיעול קדמיי, אישתאלת לרבי שמעון בן לקיש, שאלה רבי שמעון בן לקיש לרבי יוחנן, עאל רבי יוחנן ודרשה בבי מדרשא דרבי בנייה, אפילו ממזר תלמיד חכם וכהן גדול עם הארץ, ממזר תלמיד חכם קודם לכהן גדול עם הארץ... "There were two families in Sepphoris, a senatorial family and a family of commoners, who greeted the Nasi every day. The senatorial family went in first and came out

---

137 TB Yoma 78a.

138 Horayot iii, 8; and cf. Tosefta Horayot ii, 10 ...אבל אם היה הממזר תלמיד חכם וכהן גדול עם הארץ, ממזר תלמיד חכם קודם הוא לכהן גדול עם הארץ, שנאמר, "יקרה היא מפנינים" מכהן גדול הנכנס לפניי לפנים "...But if the mamzer was a talmid ḥakham and the high priest an 'am ha-aretz, the mamzer who is a talmid ḥakham takes precedence over the high priest who is an 'am ha-aretz, as it is said, 'She is more precious than inside ones' [lit. jewels], [that is,] than the high priest who enters the innermost part of the Temple, the Holy of Holies." And see Numbers Rabbah vi, 1.

a) The offspring of a union prohibited under penalty of death or of karet [divine punishment by sudden or premature death].

b) A descendant of the Gibeonites, given as hewers of wood and drawers of water for the congregation and for the altar (Jos. 9:3 ff., 27).

first. The family of commoners acquired learning in the Torah, and wanted to go in first. The question was put to R. Simeon b. Laḳish, who in turn asked R. Johanan, and R. Johanan went up and lectured in the bet ha-midrash of R. Bannayah that even in the case of a mamzer who is a talmid ḥakham and a high priest who is an ʻam ha-aretz, the mamzer who is a talmid ḥakham takes precedence over the high priest who is an ʻam ha-aretz..."[139]

The halakhah lays down that also as far as burial is concerned, the talmid ḥakham takes precedence over the ʻam ha-aretz. עיר שמתו בה שני מתים, מוציאין את הראשון ואחר כך מוציאין את השני... חכם ותלמיד, מוציאין את החכם, תלמיד ועם הארץ, מוציאין את התלמיד "A city in which there are two deceased persons, the [one who died] first is taken out for burial and then the second one is taken out for burial... [If they are] a Sage and a talmid ḥakham, the Sage is taken out [first] for burial; a talmid ḥakham and an ʻam ha-aretz, the talmid ḥakham is taken out [first] for burial."[140]

A talmid ḥakham may not forgo his rights and honour in favour of an ʻam ha-aretz, and if he does he is regarded as despising the Torah. For example, in connection with grace after meals, והא אמר רבי יוחנן, כל תלמיד חכם שמברך לפניו אפילו כהן גדול עם הארץ, אותו תלמיד חכם חייב מיתה, שנאמר, "כל משנאי אהבו מוות." אל תקרי משנאי אלא משניאי "But has not R. Johanan said, Any talmid ḥakham who allows even a high priest who is an ʻam ha-aretz to say grace before him, that talmid ḥakham commits a mortal offence, as it is said, ʻAll who hate me [mesanne'ai] love death.' Do not read mesanne'ai [who hate me] but masni'ai [who make me hated]."[141]

In connection with some subjects there occurs the expression, "It is prohibited to say this in the presence of ʻammei ha-aretz," which does not necessarily mean an halakhic discrimination against the ʻammei ha-aretz, since at times it was due to a fear that the ʻam ha-aretz would, because of his ignorance, draw the wrong conclusions. For example, אמר רבי יוחנן משום רבי שמעון בן יוחי, אפילו לא קרא אדם אלא קרית שמע שחרית וערבית קיים "לא ימוש," ודבר זה אסור לאומרו בפני עמי הארץ "R. Johanan said in the name of R. Simeon b. Yoḥai, Even though a man reads only the shemaʻ morning and evening, he has fulfilled the precept that ʻ[This book of the law] shall not depart [out of your mouth].' It

139 TJ Shabbat xiii, 13c.

140 Semaḥot xi, 1, ed. Higger, pp. 186–187.

141 TB Megillah 28a.

is however forbidden to say this in the presence of ʻammei ha-aretz.''[142] The prohibition was apparently due to educational reasons, that the ʻammei ha-aretz should not think it sufficient merely to recite the shemaʻ and that there was no obligation to study the Torah.

At times discriminating remarks against the ʻammei ha-aretz occur in the context of statements that express the hatred felt by talmidei ḥakhamim for the ʻammei ha-aretz.[143] For example, תניא, רבי אומר, עם הארץ אסור לאכול בשר (בהמה), שנאמר, ״זאת תורת הבהמה והעוף,״ כל העוסק בתורה מותר לאכול בשר בהמה ועוף, וכל שאינו עוסק בתורה אסור לאכול בשר בהמה ועוף "It was taught, Rabbi [R. Judah ha-Nasi] said, An ʻam ha-aretz may not eat the flesh of cattle, for it is said, ʻThis is the law [Torah] pertaining to beast and bird.' Whoever engages [in the study of] the Torah may eat the flesh of beast and of bird, but whoever does not engage [in the study of] the Torah may not eat the flesh of beast and of bird.''[144]

The halakhah embodies differences which applied to, and derived from, the very nature of the ʻammei ha-aretz. For example, מוכרין לעם הארץ ספרים, תפילין ומזוזות, ואין לוקחין תפילין אלא מן המומחה "One sells biblical books, tefillin, and mezuzot to an ʻam ha-aretz, but one buys tefillin only from an approved person.''[145] In tractate Miḳva'ot there are several halakhot which deal with the laws relating to an object that interposes during the immersion in a ritual bath of a person, of an article, or of clothes. In these halakhot the principle is laid down that כל המקפיד עליו חוצץ, ושאינו מקפיד עליו אינו חוצץ "If it is something which one finds annoying, it interposes, but if it is something which one does not find annoying, it does not interpose.''[146] Thus anything which a person finds annoying and does not want on his body or on his garments interposes, and anything which a person does not find annoying or disturbing does not interpose. With regard to the laws of interposition in the case of garments on which a grease stain has been found, it has been taught על הבגדים, מצד אחד אינו חוצץ, משני צדדין חוצץ. רבי יהודה אומר משום רבי ישמעאל, אף מצד אחד. רבי יוסי אומר, של בנאים מצד אחד ושל בור משני צדדין "It does not interpose [if found] on clothing on one side [only], but [if found] on both sides it interposes. R. Judah said in the name of R. Ishmael, On one side also. R. Jose said, In the case

142 TB Menaḥot 99b; the expression also occurs in TB Nedarim 49a.

143 We deal with such statements in detail below, pp. 172–180.

144 TB Pesaḥim 49b.

145 Tosefta ʻAvodah Zarah iii, 8.

146 Miḳva'ot ix, 3, 7.

of banna'im it interposes also on one side, but in the case of the uncultured, only if on both sides."[147] In a discussion in tractate Shabbat, R. Simeon b. Laḳish asked R. Ḥanina[148] whether in the case of an ass's saddle a grease stain interposed if it was on one side only, or had it to appear on both sides of the saddle. In reply the above-mentioned Mishnah from Miḳva'ot was quoted, and on its basis it was laid down ולא תהא מרדעת חשובה מבגדו של עם הארץ "Surely a saddle does not stand higher than the garment of an 'am ha-aretz."[149] There was accordingly a view of the Mishnah in Miḳva'ot which held that the terms banna'im [בנאים] and bur [בור] referred to talmidei ḥakhamim and 'ammei ha-aretz. And indeed in the continuation of the discussion in tractate Shabbat, R. Johanan said מאי בנאין — אמר רבי יוחנן, אלו תלמידי חכמים שעוסקין בבנינו של עולם כל ימיהן "What are banna'im? R. Johanan said, These are talmidei ḥakhamim who all their days engage in the upbuilding of the world." This view gives yet another halakhic difference between the 'ammei ha-aretz and the talmidei ḥakhamim which derived from differences of character, approach, and mode of life, without any intention to deny the 'ammei ha-aretz their rights.

As for the actual explanation of the Mishnah in Miḳva'ot, the view in tractate Shabbat does not, it seems, fit in with its literal meaning, while R. Johanan's interpretation is rather in the nature of an homiletical comment. It is more probable that we have here a difference between two types of artisans — the one, banna'im, builders, who were particular about the cleanliness of their clothes and hence a grease stain interposed even if on one side only, and the other, well-diggers, who were less particular about the cleanliness of their clothes and therefore a stain did not interpose unless it appeared on both sides. Even if we concede that the Mishnah refers to an uncultured person [בור], the complete identity between the terms 'am ha-aretz and an uncultured person is still not proved, so that in this instance, too, it is doubtful whether the Mishnah referred originally to 'ammei ha-aretz la-Torah.[150]

---

147 Miḳva'ot ix, 6.

148 According to the Munich MS., he asked R. Johanan.

149 TB Shabbat 114a.

150 Instead of בנאים, which occurs in our versions, the reading in the Cambridge and Kaufmann MSS. is בנווים, in which case the translation would be "with beautiful clothes." About such clothes one is particular, so that even a stain on one side only interposes. According to this reading it is however difficult to understand what בור, which is in contrast to בנווים, can be. Cf. Tosefta Miḳva'ot vi, 14; Lieberman, *Tosefet Ri'shonim*; and see Albeck, Mishnah Commentary, "Hashlamot ve-Tosafot,"

Sometimes a difference is made between the 'am ha-aretz and the ḥaver not in a decided halakhah but in a hypothetical case. For example, the halakhah in Mishnah Ketubbot lays down that if a man undertook to give a fixed sum of money to his intended son-in-law, but the son-in-law died before the wedding took place, the bride's father may say to the deceased man's brother who has to perform the levirate marriage לאחיך הייתי רוצה לתן, ולך אי אפשי לתן "I was willing to give the sum of money to your brother but I am unwilling to give it to you."[151] In this connection a Baraita in the Talmud declares תנו רבנן, אין צריך לומר ראשון תלמיד חכם ושני עם הארץ, אלא אפילו ראשון עם הארץ ושני תלמיד חכם, יכול לומר, לאחיך הייתי רוצה ליתן, לך אי אפשי ליתן "Our Rabbis taught, Where the first was a talmid ḥakham and the second an 'am ha-aretz there is no need to state that [the father-in-law] may say, 'I was willing to give [the sum of money] to your brother but I am unwilling to give it to you,' but even where the first was an 'am ha-aretz and the second a talmid ḥakham he may also say so."[152] The actual need to emphasize that there is here no difference whether the first was a talmid ḥakham and the second an 'am ha-aretz or vice versa testifies to a state of mind that in this instance the halakhah was in principle likely to make a difference between a talmid ḥakham and an 'am ha-aretz.

It should not be assumed that in the historical situation in Babylonia in the period of the Amoraim there actually existed a social stratum which was defined as the 'ammei ha-aretz. Yet in its discussions the Babylonian Talmud uses the term 'am ha-aretz when it generally refers to the concept of the 'am ha-aretz la-Torah. This it does in order to elucidate problems and resolve contradictions that arise in the course of a discussion by declaring that a certain halakhah refers to an 'am ha-aretz, while another contradictory halakhah refers to a talmid ḥakham, and so on. It is clear that in this way further halakhic differences were created between the 'ammei ha-aretz and the talmidei ḥakhamim, even though these differences were naturally in the main theoretical.[153] In general one gains the impression that the Babylonian Amoraim showed moderation as compared to the Eretz Israel Amoraim and especially

---

on this Mishnah; Dictionaries of the Talmuds; H. M. Pineles, *Darkah shel Torah*, Vienna, 1861, §143, pp. 167–168; A. Schwarz, "Ein von Resch Lakisch angedeuteter Kopistenfehler in der Mischna," *MGWJ*, 71 (1927), pp. 8–13.

151 Ketubbot vi, 2.

152 TB Ketubbot 66b.

153 See, for example, TB Nedarim 24a; TB Bava Ḳamma 95b; TB Nedarim 20a; TB Berakhot 52b.

to the Tannaim when dealing with the subject of the ʻam ha-aretz. The reason for this was of course that in Babylonia it was not such an acute subject, coupled with which was the fact that the discussions involving the concept of the ʻam ha-aretz were mainly theoretical.

7. The Development of the Concept of the ʻAm ha-Aretz

To sum up: the concept of the ʻam ha-aretz le-mitzvot, which occurred as early as in the days of the Second Temple, continued after the destruction of the Temple, was especially prevalent in the generation of Usha, and persisted until the beginning of the days of the Eretz Israel Amoraim.

The concept of the ʻam ha-aretz la-Torah occurred after the destruction of the Second Temple [unless we ascribe the saying in Avot ii, 5 to Hillel the Elder], was also particularly prevalent in the generation of Usha, and continued as an active social concept until the beginning of the period of the Eretz Israel Amoraim.

In the Introductory Remarks to this chapter [see p. 69, above] we pointed out that the concepts of the ʻam ha-aretz le-mitzvot and the ʻam ha-aretz la-Torah define essentially the same social stratum, its definition as the one or the other being dependent either on the identity of the Sage who defined and dealt with it or on the historical changes and circumstances of the periods in which the concept occurred.

At the centre of the religious world of the Pharisees were the tithes and ritual purity; the associations of ḥaverim centred round the observance of these commandments in all their minutiae and stringencies. Against this background there developed the concept of the ʻam ha-aretz le-mitzvot, a concept which referred specifically to these commandments. The ʻammei ha-aretz were the ordinary people, the rural and urban masses, who had indeed not discarded the yoke of the Torah and the commandments, but had lightened this yoke for themselves in cases in which it appeared to them to be too burdensome. These cases included the separation of the tithes and the scrupulous observance of ritual purity. This circumstance had a profound effect on the halakhah, since those scrupulous in their observance of the commandments of the tithes and of ritual purity were enjoined to shun the ʻammei ha-aretz for fear of transgressing, in the course of contact with them, through ṭevel and impurity. In addition to all this, the term ʻam ha-aretz also expressed a certain measure of disparagement and condemnation of those who were not so scrupulous in these matters.

A particularly large number of sayings and halakhot that refer to the 'am ha-aretz le-mitzvot occurred in the generation of Usha. This can be explained by the following factors which complement one another.

a) The Sages of the generation of Usha sought to describe the historical situation which obtained in the days of the Second Temple and which produced the concepts of the 'am ha-aretz le-mitzvot and of the ḥaverim who, by contrast, were scrupulous in the observance of the tithes and of purity. A description of the days of the Second Temple constituted an area in which they could well, in the periods of Jabneh and Usha, distinguish themselves, seeing that, on the one hand, there were still available old men who had been active in the days of the Second Temple or their pupils who transmitted evidences in their names, while on the other hand the members and leaders of the nation lived in the hope that "the Temple would be speedily rebuilt" and all that prevailed in its days would be restored.

b) The period of Usha represents a central and decisive stage in the redaction of the Mishnah. It is a period in which a profusion of statements and halakhot was to be found on almost every subject that occurs in the realm of the halakhah, for the Sages of the period were engaged in collecting the tannaitic material and in laying the foundation for the completion of the Mishnah in the days of R. Judah ha-Nasi. It is no wonder, then, that they also occupied themselves with the subject of the 'am ha-aretz le-mitzvot and with the related halakhot, although it is very possible, since later Sages were accustomed to transmit earlier halakhot in their own names, that some of the statements refer to an historical situation prior to their period.

c) The commandments relating to the tithes and to ritual purity persisted after the destruction of the Second Temple. Already in the period of Jabneh the leadership of the nation made every effort to preserve as much as possible the halakhot associated with Jerusalem and with the spiritual centre in order to fill the vacuum created by the destruction of the Temple. This tendency was reinforced in the period of Usha, when the grave consequences of the Bar Kokheva revolt and the subsequent religious persecution again endangered the continued existence of the world of Judaism. Assigning a central value to the religious halakhot, such as the tithes and ritual purity, that had a social significance, the Sages of Usha, in order to ensure the religious-national-social survival of the Jewish people, emphasized the importance of these halakhot and condemned as 'ammei ha-aretz those who were not scrupulous in their observance. This they did particularly in the case of commandments of the type of the tithes and ritual purity which, because of the religious persecution and the severe economic crisis, were undoubtedly suspended after the Bar Kokheva revolt. Accordingly the Sages of Usha did what they could to reinforce these commandments. It may be said that they wanted the stringencies relating to the tithes and to purity, which had been observed in the period of the Second Temple by individuals, by Pharisees, and by ḥaverim, to be observed in their days by the entire nation, including the masses and the ordinary people.

References to the concept of the ʿam ha-aretz le-mitzvot occurred until the first generations of the Eretz Israel Amoraim. In the third century CE the observance of the laws of the tithes and of ritual purity declined and the practical use of the concept of the ʿam ha-aretz le-mitzvot in reference to these commandments naturally diminished until it disappeared completely.

The concept of the ʿam ha-aretz la-Torah as an active social concept occurred, alongside the ʿam ha-aretz le-mitzvot, from after the destruction of the Second Temple until the beginning of the days of the Eretz Israel Amoraim.

The study of the Torah became the focus of Jewish life after the destruction of the Second Temple. Then it was that the leadership of the nation took the initiative not only in enlarging the value of the study of the Torah and the esteem in which its students were held, but also in abolishing sects and divisions in the nation. Thus we no longer hear of the parties of the Pharisees, the Sadducees, and the Essenes after the destruction of the Temple, nor of the sects, such as that of the Zealots, which were formed shortly before its destruction. Rejected by the nation were the sects that refused to be absorbed into it, like the various Christian as well as the gnostic sects. An effort was made to mitigate dissension in general and the conflict between Bet Hillel and Bet Shammai in particular. The associations of ḥaverim, which centred round the observance of the tithes and of purity with an extreme scrupulousness and with an excessive stringency, ceased to exist. Parties, sects, and associations were no longer the decisive form of organization. In their stead there evolved the community. In place of sects and associations, acceptance into which entailed a probationary period and was open to individuals, there came the community, in which everyone was a member and everyone was bound by a religious discipline because of his Judaism. In place of a decision by ḥaverim there was now a religious-national leadership of Sages.

This development was rendered necessary by the times, in order, as previously mentioned, to bring down the barriers separating the various sections of the people, to achieve as far as possible the equality of the different circles in the nation, and to find a purpose that would fill the vacuum created by the destruction of the Temple. Before its destruction, the restrictions relating to purity were abolished specifically during a pilgrim festival, when it was laid down that all the people were like ḥaverim, in order that the unity of the nation might be preserved. Even so, when the Temple which had united the nation around it had been destroyed, there was a need to find ways to sustain the unity of the nation and to create new aims and objectives while maintaining, to the greatest possible extent, the commandments associated with the Temple and with Jerusalem. The factor which united the nation and the aim and purpose which were set before every Jew lay in the realm of the study of the Torah and of the esteem accorded to its students. Clearly therefore, if the talmid ḥakham climbed to the topmost rung of the social ladder, the 'am ha-aretz was placed at its lowest rung. His ignorance, his lack of knowledge of the Torah, and his shunning the company of talmidei ḥakhamim, all this was expressed by the derogatory and contemptuous term of 'am ha-aretz.

The concept of the 'am ha-aretz la-Torah was also extremely prevalent in the period of Usha, and this, too, was clearly the result of the tendency of the Sages to reinstate the process of setting the study of the Torah and its students at the centre of Jewish life, a process which was liable to be undermined and to collapse as a result of the religious persecution, the economic difficulties, and the migration to Galilee. Accordingly it becomes clear why this period included the astonishingly acrimonious statements about the 'am ha-aretz la-Torah and exhibited a very deep gap between the 'ammei ha-aretz and the talmidei ḥakhamim, subjects we shall later deal with at length.

As an active social term reflecting the contemporary historical situation the concept of the 'am ha-aretz la-Torah continued up to the early days of the Eretz Israel Amoraim. In the third century CE it passed more and more into disuse, since the 'am ha-aretz was recognized as a healthy component in the nation alongside the social elite of the type of the talmidei ḥakhamim, and as capable of being meritorious in the sphere, for example, of moral values and virtues.[154]

---

154 We revert to this below, pp. 188–195.

CHAPTER FOUR

THE 'AMMEI HA-ARETZ, THE PHARISEES, AND THE ḤAVERIM

1. Introductory Remarks

The 'am ha-aretz le-mitzvot is contrasted in talmudic literature with the Pharisee, and in numerous halakhot and sayings with the ḥaver in particular. In observing the commandments and restrictions relating to tithes and to purity the Pharisees and the ḥaverim were scrupulous, contrary to the 'ammei ha-aretz who disregarded their observance. The strict observance by the former, and the disregard by the latter, of the separation of tithes, and more especially of the various degrees of purity, led to a social division between the Pharisees and the ḥaverim on the one hand and the 'ammei ha-aretz on the other.

A distinction between the social strata based on the observance of purity occurs in an ancient Mishnah which we have had occasion to quote several times. בגדי עם הארץ מדרס לפרושין, בגדי פרושין מדרס לאוכלי תרומה, בגדי אוכלי תרומה מדרס לקדש, בגדי קדש מדרס לחטאת וכו׳ "The garments of an 'am ha-aretz are a source of midras-impurity to Pharisees, the garments of Pharisees are a source of midras-impurity to those who eat terumah, the garments of those who eat terumah are a soucre of midras-impurity to those who eat holy things, the garments of those who eat holy things are a source of midras-impurity to those who attend to the water of purification..."[1] Here a distinction is made between different strata — between the 'ammei ha-aretz, the Pharisees, those who eat terumah, and so on — the criterion for the distinction being the degree to which the laws of purity were observed.

2. The Ḥaverim and the Association

It is in particular with the ḥaver that the 'am ha-aretz was contrasted. By the nature of things, the ḥaverim were very close to the Pharisees in that the latter were scrupulous about the same areas in which the

[1] Ḥagigah ii, 7.

ḥaverim adopted restrictive practices. But whereas the Pharisees constituted a spiritual-social movement, the ḥaverim belonged to closed associations. This can be seen from the conditions governing admission to these associations, from the obligations binding on the ḥaverim in the associations, and from the customs prevailing in them, as also from other factors, all of which combine to present a picture reminiscent of the life of a fraternal order.

In talmudic literature there are collections somewhat in the nature of codes which contain the laws of these associations and the conditions of admission to them. Such halakhic collections occur in the Mishnah, in the Tosefta, in TJ Demai, and in TB Bekhorot.[2] In addition to these there are numerous halakhot and statements scattered throughout talmudic literature, in the vast majority of which the ḥaver and the 'am ha-aretz are contrasted with each other.

As usual in every closed association, those wishing to join it had to comply with the conditions of admission and had to undergo a probationary period.[3]

A person desirous of joining the association of ḥaverim is defined in the sources as המקבל עליו "one who takes upon himself"; and one who takes upon himself the conditions of admission and binds himself to keep the obligations of a ḥaver in the association מקבלין אותו להיות חבר "is accepted as a ḥaver."[4] In talmudic literature this expression denotes

[2] Demai ii, 2–3; Tosefta Demai ii, 2–iii, 15; TJ Demai ii, 22d–23a; TB Bekhorot 30b–31a.

[3] For the relevant sources and bibliography, see below, notes 93–95 in this chapter [pp. 147–148].

[4] The expression לקבל עליו "to take upon himself" occurs in the sources in several different connections: המקבל עליו להיות חבר "One who takes upon himself to be a ḥaver" (Demai ii, 3); המקבל עליו ארבעה דברים מקבלין אותו להיות חבר "One who takes upon himself four things is accepted as a ḥaver" (Tosefta Demai ii, 2); עם הארץ שקיבל עליו דברי חבירות "An 'am ha-aretz who took upon himself the obligations of a ḥaver" (Tosefta Demai ii, 3, 5); הבא לקבל עליו "One who comes to take upon himself" (Tosefta Demai ii, 10); המקבל עליו בפני חבורה "One who takes upon himself [the obligations of a ḥaver] before an association" (Tosefta Demai ii, 14); תני כל הבא צריך לקבל עליו "It has been taught, Whoever comes [to apply to be a ḥaver] has to take upon himself" (TJ Demai ii, 22d); תנו רבנן הבא לקבל דברי חבירות "Our Rabbis taught, One who comes to take [upon himself] the obligations of membership in an association" (TB Bekhorot 30b); etc.

at times a vow,[5] at others an oath.[6] It is consequently difficult to know whether the reference in the present case is to an obligation taken in the form of a vow or of an oath. In any event the request to be accepted into the association entailed one of these given in a public declaration in the presence of the entire association or of its representatives.[7]

After the association had decided in principle to accept the candidate, he had to undergo a probationary period which necessitated his passing through a number of stages until his final admission. There was a difference in the stages of admission of one who had before his acceptance observed the customs of the association on his own and one who had not previously done so. The latter had accordingly first to undergo a period of learning the customs and obligations of the association.[8] There

---

5 So, for example, הנושא נשים בעברה פסול עד שידיר הניה, המטמא למתים פסול עד שיקבל עליו, שלא יהא מטמא למתים "A priest who contracts an illegal marriage is unfit [for the priesthood] until he vows not to derive any benefit from the woman [that is, until he divorces her]. One who defiles himself through contact with the dead is unfit, until he undertakes not to defile himself any more through the dead" (Bekhorot vii, 7). Perhaps it is in this spirit that we are to understand the statement נדרים סיג לפרישות "Vows are a fence to abstinence" (in the name of R. Akiva; Avot iii, 13). The concept of abstinence is closely connected with the concepts of purity and holiness, round which the idea of the ḥaverim and the association centred.

6 TJ Soṭah ii, 18b; TB Shevu'ot 36a; etc. The expression in Est. 9:27 is also explained in this way (and see Tosefta Soṭah vii, 4–5). It should be pointed out that the candidate for admission to the sect of the Yaḥad [the community of the Dead Sea sect] was obliged to take the "oath of the covenant" [שבועת ברית], although this oath was of a general nature and referred to the observance of the commandments (*Damascus Document*, xv, 7–11; cf. *Manual of Discipline*, vi, 13–15).

7 Tosefta Demai ii, 14 refers to המקבל עליו בפני חבורה "One who takes upon himself [the obligations of a ḥaver] before an association," that is, the declaration had to be made before the association as a whole, whereas according to a Baraita in TB Bekhorot 30b it had to be made before three members of the association only: הבא לקבל דברי חבירות צריך לקבל בפני שלשה חבירים "One who comes to take [upon himself] the obligations of membership in an association is required to do so in the presence of three ḥaverim." This difference may have been due to a development that occurred in the laws of the associations, to differences between one association and another and one place and another, or to one of these halakhot dating from a time when the associations no longer existed, so that it fails to reflect the prevailing situation. S. Lieberman, "The Discipline in the So-Called Dead Sea Manual of Discipline," *JBL*, 71 (1952), pp. 199–206, suggests that the Baraita in Bekhorot may follow the principle that in general three represented the whole association, as, for example, שלשה מבית הכנסת כבית הכנסת "Three [members] of the synagogue are like the synagogue" (TJ Megillah iii, 74a).

8 הבא לקבל עליו אם היה נוהג כתחילה בצינעא, מקבלין אותו, ואם לאו אין מקבלין אותו ואחר כך מקבלין אותו [כ״י וינה ודפוסים: אם היה נוהג מתחילה בצינעה, מקבלין אותו

was however no difference in the conditions governing the admission of one who was, and one who was not, a talmid ḥakham. הבא לקבל עליו אפילו תלמיד חכם צריך לקבל עליו "One who comes to take upon himself [the obligations of a ḥaver], even a talmid ḥakham has to take upon himself [the obligations of a ḥaver]."[9] Similarly one who was an 'am ha-aretz could be accepted into the association: עם הארץ שקיבל עליו דברי חבירות "an 'am ha-aretz who took upon himself the obligations of a ḥaver."[10]

The probationary period for membership in the association involved not only learning its customs and ways but also additional stages before the final admission. והולכין ומקבלין לכנפים ואחר כך מקבלין אותו לטהרות. ואם אינו מקבל עליו אלא לכנפים בלבד מקבלין אתו. קיבל עליו לטהרות ולא קיבל עליו לכנפים, אף על הטהרות אינו נאמן. עד מתי מקבלין, בית שמאי אומרים למשקין שלשים יום ולכסות שנים עשר חודש, ובית הילל אומרים זה וזה שלשים יום "They proceed to accept him if he undertakes to observe cleanness of hands [kenafayim], and afterwards he is accepted for the observance of the laws of purity. If he takes upon himself the observance of the cleanness of hands alone, he is accepted. If he takes upon himself the observance of the laws of purity but has not taken upon himself the observance of the cleanness of hands, he is not deemed to be a ne'eman[a] even with regard to the laws of purity. How long is the period that has to elapse before he is accepted as a ḥaver? Bet Shammai said, As regards the purity of liquids, the period is thirty days, and as regards the purity of his garments, the period is twelve months, whereas Bet Hillel said, In both cases the period is thirty days."[11]

ואחר כך מלמדין אותו, ואם לאו, מלמדין אותו ואחר כך מקבלין אותו]. ר' שמעון אומר, לעולם מקבלין אתו ומלמדין אתו "One who comes to take upon himself [the obligations of a ḥaver], if he previously observed [them] in private, he is accepted, and if not, he is not accepted and afterwards he is accepted [Vienna MS. and printed editions: 'if he previously observed [them] in private, he is accepted and afterwards taught, but if not, he is taught and afterwards accepted']. R. Simeon said, He is always accepted and [then] taught" (Tosefta Demai ii, 10).

9 Tosefta Demai ii, 13; and parallel passages. For different views on this halakhah, see below, p. 144.

10 Tosefta Demai ii, 5; and parallel passages. On this, see below, pp. 163–164.

11 Tosefta Demai ii, 11–12. However TB Bekhorot 30b has בית הלל אומרים אחד זה ואחד זה לשנים עשר חודש "Bet Hillel said, In both cases the period is twelve months." On this the Talmud (*ibid.*) raises the question אם כן הוה ליה מקולי בית שמאי ומחומרי בית הלל, אלא בית הלל אומרים אחד זה ואחד זה לשלשים "If this is so, then we have here a ruling in which Bet Shammai adopt the more lenient and Bet

a) Lit. "trustworthy." On this term, see below, pp. 151–156.

The term kenafayim [כנפים] apparently refers, as translated above, to cleanness of hands.[12] On the extent of the purity required of the candidate at this stage there were differences of opinion as early as among the Eretz Israel Amoraim. תני, מקריבין לכנפים ואח״כ מלמדים לטהרות. אמר ר׳ יצחק בר׳ אלעזר, כנפים מדפות, טהרות היסטות. ר׳ מנא אמר, כנפים מדפות היסטות. טהרות מעשרות. "It has been taught, He is accepted if he takes upon himself to observe cleanness of hands, and afterwards he is taught to observe the laws of purity. R. Isaac b. Eleazar said, He takes upon

Hillel the stricter view. Read rather, Bet Hillel said, In both cases the period is thirty days." The version in the Babylonian Talmud is probably the original one, the halakhah in the Tosefta having been emended so as not to make it a case of the leniencies of Bet Shammai and the restrictions of Bet Hillel.

12 According to the halakhah ...וחכמים אומרים את שנטמא באב הטמאה מטמא את הידים, בולד הטמאה אינו מטמא את הידים "...But the Sages said, That which has been rendered impure by 'a father of impurity' conveys uncleanness to the hands, but that which has been rendered impure by 'an offspring of impurity' does not convey uncleanness to the hands" (Yadayim iii, 1). And cf. תורת הנזיר בין שיש לו כנפים בין שאין לו כנפים " 'This is the law for the Nazirite,' denotes whether he has hands or not" (TJ Nazir vi, 55c); in the parallel passage in the Babylonian Talmud (TB Nazir 46b) the reading is כפיים instead of כנפים (and see *Aruch Completum*, as well as the other dictionaries of the Talmuds).

Already R. Gershom Me'or ha-Golah explained the term in this manner in his commentary to Bekhorot 30b.

Several scholars, some of whose remarks have a bearing on the nature of the stages in the process of admission to the association, have given different interpretations to the term.

According to N. Brüll, "The Basis and Development of the Laws relating to the Purity of the Hands" (Hebrew), *Bet Talmud*, II (1881), pp. 368 ff., כנפים signifies tefillin. Brüll bases his view on the fact that in the halakhah כנפים and purity relate to separate stages. There is consequently no connection between the two. The relation between כנפים [= wings, generally] and tefillin derives from the shape of the tefillin and their straps. In support of his view he quotes the statement אמר רבי ינאי תפילין צריכין גוף נקי, כאלישע בעל כנפים "R. Yannai said, Tefillin require a pure body, like Elisha, the man of כנפים, of wings" (TB Shabbat 49a; and see TJ Berakhot ii, 4c). Brüll connects this with the view איזהו עם הארץ, כל שאינו מניח תפילין "Who is an 'am ha-aretz? Anyone who does not put on tefillin" (TB Berakhot 47b; and cf. Tosefta Demai ii, 17). Hence, according to Brüll, one of the conditions for admission to the association was the putting on of tefillin.

Luzzatto and, following him, A. Büchler, *Der galiläische 'Am-ha'Areṣ des zweiten Jahrhunderts*, Vienna, 1906, pp. 168–170, explain כנפים as the skirts of a garment (Hag. 2:12; and cf. Sifra xii, 13, ed. Weiss, 27a). According to them, one of the stages in the probationary period required of the candidate that he be careful not to have his garments defiled by his wife's impurity. This corresponds to the grading in Tosefta Ṭohorot iii, 9–10: התחיל יוצא ונכנס בגדיו טהורין, ואין עושין על גביו טהרות... היה יודע לשמור את גופו אוכלין על גב גופו טהרות, לשמור את ידיו אוכלין על גב ידיו

himself to observe cleanness of hands should they come in contact with maddafot,[a)] and to observe the laws of purity as far as hesseṭot[b)] are concerned. R. Mana said, Cleanness of hands has to be observed in the case of both maddafot and hesseṭot, the laws of purity in the case of tithes."[13]

According to the view of R. Isaac b. Eleazar, the stages clearly set forth in the halakhah in the Tosefta are as follows.

1. *He is accepted for the observance of the cleanness of hands*

The ancient halakhah lays down that hands which come in contact with impurity in the first degree are rendered impure in the second degree.[14] Maddafot are contacts made with "a father of impurity" [אב הטומאה] and are in the first degree of impurity,[15] so that hands which come in contact with maddafot are impure and have to be

---

טהורות "If he [the child] has begun to go in and out, his clothes are pure. Nevertheless, secular food is not prepared in purity for him... If he knows how to look after his body, secular food prepared in purity is eaten on account of [the purity of] his body. If he knows how to look after his hands, secular food prepared in purity is eaten on account of [the purity of] his hands."

S. Krauss, "חבר עיר. Ein Kapitel aus altjüdischer Kommunalverfassung," *JJLG*, 17 (1926), pp. 195–241, maintains that כנפים is derived from the Aramaic root כנף, which occurs in particular in the word כנופיה [= an assembly], the relation between כנף and כנפים being, according to Krauss, like that between כנס and כנסת. In אלישע בעל כנפים "Elisha, the man of כנפים," the reference is to Elisha who belonged to the association, this being an expression analogous to בני הכנסת "the members of the synagogue" (Zavim iii, 2). According to this, מקבלין לכנפים means that he is admitted to the assembly or to the association.

For other explanations, see H. Graetz, *Geschichte der Juden*, Leipzig, 1906[5], pp. 699–702, note 12; Simhoni, Notes to the Hebrew translation of *The Jewish War*, entitled *Milḥamot ha-Yehudim*, ii, chap. 8, p. 424; L. Herzfeld, *Geschichte des Volkes Jisrael*, Leipzig, 1863, II, p. 391; etc.

To sum up: the term is undoubtedly obscure, but according to the context the most probable explanation is that given here, namely, that it refers to the cleanness of the hands. [See Lieberman, *JBL*, 71 (1952), pp. 199–206; and his commentary to Tosefta Demai on the above-mentioned halakhot.]

13 TJ Demai ii, 23a, according to the version of the Rome and the Solomon Sirillo MSS.; and see Lieberman, *Tosefta ki-Peshuṭah*, Seder Zera'im, p. 215.

14 See above, p. 56.

15 תולדות השרץ אי זהו, ר׳ יודה בשם ר׳ נחום: מדפות, מהו מדפות, מגעות "What are the offsprings of an unclean reptile? R. Judah in the name of R. Nahum said, Maddafot. What kind of maddafot? Contacts" (TJ Shabbat vii, 9d).

a) For an explanation, see paragraph 1 that follows immediately.

b) For an explanation, see paragraph 2 that follows immediately after paragraph 1.

washed to render them pure once more.[16] Accordingly, at the stage of cleanness of hands the candidate took upon himself to wash his hands should they be rendered impure by maddafot.

2. *He is accepted for the observance of the laws of purity*

At this stage the candidate took upon himself to exercise care even in the case of hesseṭot. Hesseṭot refer to the impurity conveyed to an object which is shaken by an impure person without his touching or carrying it; or to the impurity contracted by a pure person through an impure object which is shaken by him without his touching or carrying it. In the former case a lenient view is adopted, the halakhah stating that hesseṭot of this kind do not render secular things impure.[17] At the second stage of his probationary period the candidate had to guard himself against impurity, even that caused by hesseṭ, and if he was defiled through such an impurity, he was to purify himself, even for the eating of secular food. Thus this second stage refers to additional restrictions which the association took upon itself in the domain of purity.

According to R. Mana's view of the halakhah in the Tosefta, already at the first stage the candidate took upon himself to be careful as regards both maddafot and hesseṭot, since according to him hesseṭot also entailed the cleanness of hands. At the second stage the candidate took upon himself to observe the purity of the tithes, which apparently refers to the eating of secular food in the state of purity required for tithes, this being of course also to some extent an additional restriction.

Further stages in the process of admission to the association are to be found in the dispute between Bet Shammai and Bet Hillel in the Tosefta. Without our going into details, it is evident from this dispute that during the various stages of the probationary period the candidate attained a degree at which he was held to be trustworthy as regards the purity of liquids, and also a degree at which he was deemed to be trustworthy as regards the purity of his garments.

---

16 Cf. Tosefta Yadayim ii, 9: אלו פוסלין לידים, אוכל ראשון והמדפות והמשקין "These make the hands unfit, food in the first degree of impurity, maddafot, and liquids." See also Yadayim ii, 1.

17 See Zavim iii, 2: וכלן טהורין לבני הכנסת וטמאין לתרומה "In both cases however they are pure for members of the synagogue, but impure for terumah"; and see also Ṭohorot vii, 5, where it says that המניח עם הארץ בתוך ביתו לשמרו "if a man left an 'am ha-aretz in his house to guard it," כלי החרס המקפין צמיד פתיל טהורין "earthenware that has tightly fitting covers remains pure"; and cf. also TJ Ḥagigah ii, 78c.

It is clear from this halakhah that there existed a view according to which the candidate was regarded as trustworthy in respect of the purity first of liquids and then of garments. This is a surprising distinction, since in matters relating to purity the halakhah as a rule applies the stricter view specifically to liquids, as evidence of which there is the principle משקין תחילה לעולם "Liquids are always rendered impure in the first degree."[18] It may however be that a more lenient view was adopted as regards liquids since their impurity is a rabbinical enactment.[19] The importance attached to the purity and impurity of garments is evident from the fact that it was one of the criteria which defined the difference, in the realm of purity, between various strata of people, as, for example, in Ḥagigah ii, 7, a Mishnah we have previously quoted several times.[20]

It may be that the distinction here between liquids and garments is not due to the relative severity of their impurity, but refers rather to stages in the establishment of closer relations between the association and the candidate, in that at first the ḥaverim ate with him and of his food without fear of impurity, and afterwards they were guests of his and he was a guest of theirs while he wore his own clothes.[21]

Several passages set forth the conditions of admission to the association.

In Mishnah Demai: המקבל עליו להיות חבר, אינו מוכר לעם הארץ לח ויבש, ואינו לוקח ממנו לח, ואינו מתארח אצל עם הארץ, ולא מארחו אצלו בכסותו. רבי יהודה אומר, אף לא יגדל בהמה דקה, ולא יהא פרוץ בנדרים ובשחוק, ולא יהא מטמא למתים, ומשמש בבית המדרש. אמרו לו, לא באו אלו לכלל "One who takes upon himself to be a ḥaver may not sell to an 'am ha-aretz moist or dry [produce], nor may he buy from him moist [produce]. He may not be the guest of an 'am ha-aretz, nor may he receive as a guest an 'am ha-aretz who is wearing his own clothes. R. Judah said, He may also not breed small cattle, nor may he be addicted to making vows or to laughter, nor may he defile himself with the dead, but he must be an attendant at the house of study. They however said to him, These [requirements] do not come within the general rules [governing membership in an association]."[22]

---

18 See above, pp. 60–61.

19 See TB Pesaḥim 16b. And see Rashi on the Baraita in TB Bekhorot 30b.

20 See also Ṭohorot iv, 5; Tosefta Ṭohorot v, 16; TB Ḥullin 35b.

21 Cf. Demai ii, 3, where it is stated that a ḥaver should keep aloof from an 'am ha-aretz wearing his own clothes.

22 Demai ii, 3.

In Tosefta Demai: המקבל עליו ארבעה דברים מקבלין אותו להיות חבר, שלא יתן תרומה ומעשרות לעם הארץ, ושלא יעשה טהרותיו אצל עם הארץ, ושיהא אוכל חוליו בטהרה "He who takes upon himself four things is accepted as a ḥaver: that he will not give terumah and tithes to an ‘am ha-aretz, that he will not prepare his food in the observance of the laws of ritual purity with an ‘am ha-aretz, and that he will eat his secular food in a state of purity."[23]

Although the halakhah in the Tosefta mentions at the outset "four things" which constitute the obligations of a ḥaver, in specifying them only three are mentioned. The most probable view seems to be that of Büchler and Lieberman[24] who hold that the continuation of the halakhah in the Tosefta המקבל עליו להיות נאמן וכו׳ "who takes upon himself to be a ne'eman..." is the fourth obligation, and that the definite article should be replaced by a conjunctive "vav," so that the reading should be ומקבל עליו להיות נאמן, מעשר את שהוא אוכל ואת שהוא מוכר ואת שהוא לוקח, ואינו מתארח אצל עם הארץ "and takes upon himself to be a ne'eman, tithes what he eats and what he sells and what he buys, and is not the guest of an ‘am ha-aretz."[25]

A further source in Avot de-R. Nathan states כל המקבל עליו ארבעה דברים מקבלין אותו להיות חבר. אינו הולך לבית הקברות, ואינו מגדל בהמה דקה, ואינו נותן תרומה ומעשר לכהן עם הארץ, ואינו עושה טהרות אצל עם הארץ, ואוכל חולין בטהרה "Whoever takes upon himself four things is accepted as a ḥaver: that he does not go to a cemetery, does not breed small cattle, nor give terumah and tithes to a priest who is an ‘am ha-aretz, that he does not prepare food in the observance of the laws of purity with an ‘am ha-aretz, and that he eats secular food in the observance of the laws of purity."[26]

---

23 Tosefta Demai ii, 2. The Vienna MS. has the reading ושלא יעשה טהרות לעם הארץ "And that he will not prepare food in the observance of the laws of purity for an ‘am ha-aretz," but this is unacceptable. According to Lieberman, *Tosefta ki-Peshuṭah*, Seder Zera‘im, p. 210, this version is due to a scribal error because of Tosefta Demai iii, 1 which has such a reading. In the Vienna MS. and in published versions the reading is ושיהא אוכל חולין בטהרה "And that he will eat secular food in a state of purity" [see above, p. 9].

24 Büchler, *Der galiläische ‘Am-ha'Areṣ*, p. 151; Lieberman, *Tosefet Ri'shonim*, I, p. 64; *idem*, *Tosefta ki-Peshuṭah*, Seder Zera‘im, p. 210.

25 Tosefta Demai ii, 2.

26 Avot de-R. Nathan, Version A, xli, ed. Schechter, p. 132. And see below, note 28 [the note after the next one].

Of the views contained in these sources on the duties obligatory on one wishing to be admitted to the association, the statement of the Tanna mentioned first in the Mishnah as also the halakhah in the Tosefta seem to reflect, to a greater degree than the others, the situation that prevailed in the associations. Here we once more see that the association centred round the strict observance of purity coupled with the scrupulous separation of the tithes, both of which involved shunning the 'ammei ha-aretz, as mentioned in the halakhot.

R. Judah's statement in the Mishnah appears to have been directed more at those who in his day sought to follow in the paths of abstinence and purity and to minister to talmidei ḥakhamim [שימוש תלמידי חכמים] than that it reflects the state of affairs that obtained in the associations, whose main period fell in the days of the Second Temple [as we shall see later]. This is emphasized by the prohibition against rearing small cattle which, according to R. Judah, was included among the obligations of a ḥaver, but which was instituted after the destruction of the Second Temple, during the period of Jabneh, to solve problems that had arisen because of the situation at that time.[27] So too as regards ministering to talmidei ḥakhamim, which is a matter that concerns the relations between pupils and their teacher but had no connection with the associations, for whom the observance of purity and the tithes were the central factors. The Mishnah itself adds that R. Judah was told, "These [requirements] do not come within the general rules [governing membership in an association]," which shows that such matters had nothing to do with the obligations and rules of an association.

The Baraita in Avot de-R. Nathan is also apparently a later version which combines elements not only of the statement by the Tanna mentioned first in the Mishnah but also of the Tosefta and of the remarks of R. Judah. Proof of this can be seen in the opening words of the halakhah in Avot de-R. Nathan, "Whoever takes upon himself four things is accepted as a ḥaver," which corresponds to the beginning of the Tosefta but is followed by the mention of five things.[28]

Some scholars have sought to embody the stages of admission in a unified structure by combining, either in part or in whole, the conditions and stages of admission as well as the obligations which, as set

---

[27] See G. Allon, *Toledot ha-Yehudim be-Eretz Yisra'el bi-Teḳufat ha-Mishnah ve-ha-Talmud*, Tel Aviv, I: 1958[3], II: 1961[2] — see I, pp. 173–178.

[28] See also Schechter, *op. cit.*, note 18, who found differences on the subject between the current printed version and the versions in MSS. and in older books.

forth in the halakhic sources, a candidate desirous of joining the association had to take upon himself.

Thus Lieberman,[29] in his commentary to the Tosefta, finds in halakhot 11 and 12 of Tosefta Demai ii, quoted above, the following order in the stages of admission.

a) The observance of the obligations of a ḥaver outside the organization.

b) Acceptance for the observance of the cleanness of hands.

c) Acceptance for the observance of the laws of purity.

d) Thirty days after admission, the candidate takes upon himself the obligations of a ḥaver to maintain the purity of liquids.

e) Twelve months after admission, full membership in the association [according to the view of Bet Shammai].

Rabin[30] does not consider that halakhot 11 and 12 in Tosefta Demai ii contain the various stages of admission. Instead, adopting the view that "kenafayim" means the skirt of a garment, he identifies "kenafayim" in halakhah 11 with "kesut" [= garment] in halakhah 12, and similarly connects acceptance for the observance of the laws of purity in halakhah 11 with acceptance for the observance of the purity of liquids in halakhah 12. In consequence of this, he changes the order in halakhah 12, maintaining as he does that acceptance for the observance of the purity of liquids came after acceptance for the observance of the purity of garments. This also follows from the fact that liquids are rendered impure in a special manner, and hence, in Rabin's view, it cannot be assumed that being a ne'eman in respect of liquids preceded being a ne'eman in respect of garments. Accordingly in the two halakhot the first stage is identical, namely, acceptance for the purity of garments, while the second stage begins with acceptance for the observance of the laws of purity as in halakhah 11, and ends with acceptance for the observance of the purity of liquids as in halakhah 12. The dispute between Bet Shammai and Bet Hillel centred on the duration of these two stages, namely, that of the observance of the purity of garments and that of the observance of the purity of food and liquids.

Neusner[31] coordinates the conditions of admission in halakhot 11 and 12 in Tosefta Demai ii with the halakhot in Tosefta Demai ii, 2

---

29 *Tosefta ki-Peshuṭah*, Seder Zera'im, p. 216; and see *idem*, *JBL*, 71 (1952), pp. 199–206.

30 Ch. Rabin, *Qumran Studies*, Oxford, 1957, pp. 1–21.

31 J. Neusner, "The Fellowship (חבורה) in the Second Jewish Commonwealth," *HTR*, 53 (1960), pp. 125–142.

and in Mishnah Demai ii, 3 respectively, and combines them into a unified arrangement of the stages of admission, as follows.

Tosefta Demai ii, 2	*Tosefta Demai ii, 11*
a) He will not give terumah and tithes to an ʻam ha-aretz.	Observance of the laws of purity.
b) He will not prepare his food in the observance of the laws of purity with an ʻam ha-aretz.	Observance of the laws of purity.
c) He will eat his secular food in a state of purity.	Observance of the cleanness of hands.[32]
Mishnah Demai ii, 3	*Tosefta Demai ii, 12*
a) He may not sell to an ʻam ha-aretz moist or dry [produce].	Observance of the purity of liquids.
b) He may not buy from him moist [produce].	Observance of the purity of liquids.
c) He may not be the guest of an ʻam ha-aretz.	Observance of the purity of garments.
d) Nor may he receive as a guest an ʻam ha-aretz who is wearing his own clothes.	Observance of the purity of garments.

Neusner thus sees in the probationary period two stages, each subdivided into two. The first stage is contained in halakhot 2 and 11 in the Tosefta, and its sub-stages are the observance of the purity of food and the eating of secular food in the observance of the rules of purity. On the basis of this the candidate was accepted for the observance of the laws of purity and of the cleanness of the hands. The second stage is given in the Mishnah and in halakhah 12 in the Tosefta, on the basis of which the candidate was accepted for the observance of the purity of liquids and of garments.[33]

In discussing the above-mentioned halakhot these and other scholars have said that they indicate that admission to the association was carried out in stages and was graded, as was also the case with other Jewish

---

32 Neusner accepts the interpretation of כנפים as כפים, that is, the washing of hands.

33 According to both Rabin and Neusner there was a probationary stage which preceded all the stages of admission mentioned here, namely, that of a ne'eman. On this, see below, pp. 152–156.

and non-Jewish associations at that time.[34] While we accept this view it is nevertheless doubtful whether it is at all possible to deduce from these halakhot the exact process and stages of admission, and hence the above attempts to do so appear to be rather in the nature of guesswork. It is very likely that what we have here are tannaitic sources of different periods which describe the process of admission to an association and some of which were composed or formulated a considerable time after the associations had ceased to exist. Even as already the Eretz Israel Amoraim were not clear, as we have seen, about the meaning of the term "kenafayim," so also some of the halakhot under discussion undoubtedly include traditions that are somewhat hazy and do not provide a detailed and precise description of the process of admission. Nor does it necessarily follow that all the halakhot cited deal only with the various stages in the process of admission, for those in Mishnah Demai and in Tosefta Demai ii, 2 apparently also contain the obligations of a ḥaver not just during the process of his admission but also when he was a full member of the association.

In reality no great historic importance attaches to the precise process of admission, but rather to its actual existence, a fact established beyond doubt by the sources under discussion. The scholars who have endeavoured to describe the precise process of admission as a whole have been prompted by a desire to compare the process, forced though such a comparison might be, with the stages of admission to the Essenes and the Dead Sea sect of the Yaḥad. [For these attempts, see pp. 147–151, below.]

It may be conjectured that in general the association applied the degrees of impurity in the Pharisaic halakhah to the process of admission to the association. In conformity with these degrees of impurity it may be possible to describe the stages of admission without fixing their duration or making them coincide fully with the halakhot in the Mishnah and in Tosefta Demai, as follows.

a) A person outside the association who was suspected of not being generally scrupulous in matters of purity was regarded as "a father of impurity," like one with a discharge, and was defined as an 'am ha-aretz.

b) A person wishing to be admitted to the association was, at the beginning of his probationary period, regarded as being in the first

---

[34] See below, notes 93–95 in this chapter [pp. 147–148] for the relevant sources and bibliography.

degree of impurity, and accordingly rendered secular food impure and was prohibited from touching the food of the association prepared in accordance with the laws of purity.

c) A person wishing to be admitted to the association was during the course of his probationary period regarded as being in the second degree of impurity and was therefore no longer able to render food and utensils impure, and was permitted to touch the association's food prepared in accordance with the laws of purity, but was still able to defile liquids, since השני מטמא את המשקין לעשות תחילה "the second degree of impurity renders liquids impure in the first degree,"[35] and hence he was prohibited from touching the liquids of the association.

d) When he became a member of the association he was considered pure.

According to Allon,[36] the areas in which the associations were scrupulous refer to commandments that were a definite obligation on the nation as a whole and did not represent additional restrictions taken by the associations upon themselves. The main purpose of forming the associations was, so Allon contends, to make it easier for the ḥaverim in an association to observe the commandments relating to food, in that every ḥaver would be surrounded by those who were particular about such commandments, and he would therefore have no need to deal with non-ḥaverim or with 'ammei ha-aretz who were suspected of not observing these commandments and were likely to cause the ḥaver to transgress.

It would appear that the actual structure of the association and the mechanism reflected in the gradual process of admission to it, as also its institutions and customs, with which we shall deal later, attest to the fact that the association was not merely a society whose purpose was to shun the 'ammei ha-aretz so as not to transgress with ṭevel,[a] demai, or impurity, nor was the mode of life in the association compatible with the ways of people seeking an easy means of observing the commandments of the Torah.

An analysis moreover of the obligations of a ḥaver in the association clearly shows that it was a society that took upon itself additional restrictions in the areas in which it was scrupulous.

---

35 Tosefta Ṭohorot i, 6.

36 G. Allon, *Meḥḳarim be-Toledot Yisra'el*, Tel Aviv, 1957–1958, I, pp. 148–176.

a) Produce at the stage at which tithes and terumot should be, but have not yet been, separated.

The very conditions of admission to the association, discussed above, included additional restrictions which went much beyond what was demanded in the usual observance of the commandments.

There was an additional restriction in the halakhah which prescribed that a ḥaver לא יתן תרומה ומעשרות לעם הארץ "shall not give terumah and tithes to an ʻam ha-aretz,"[37] the aim of which was to preclude the possibility of a priest who was an ʻam ha-aretz eating terumah and tithes while in a state of impurity. In this instance, since no Jew is necessarily responsible for the deeds of a fellow-Jew,[38] the ḥaverim acted beyond what was required of them. A heightened value was attached to purity, which was thus included among the essential principles of Judaism for which כל ישראל ערבים זה בזה "all Jews are responsible for one another."[39]

There was likewise an additional restriction in the prohibition imposed on the members of the association against selling to an ʻam ha-aretz moist or dry produce.[40] Here it was prohibited for a ḥaver or for a candidate desirous of joining the association to give to an ʻam ha-aretz dry food, although it is basically not susceptible to impurity,[41] for fear that it might be rendered liable to impurity while in the possession of the ʻam ha-aretz and would become defiled by his impurity. Here, too, there was imposed on the seller a concern for what might happen to the food while in the possession of the ʻam ha-aretz even in the case of secular food.

The crowning obligation of a ḥaver in the association was the eating of secular food in the observance of the rules of purity,[42] which is also to be seen as an additional restriction.

In the period of the Tannaim the eating of secular food in the observance of the laws of purity was regarded as a special virtue, so that often, when they wished to praise a Sage for his singularly good qualities, they mentioned the fact that he ate secular food in a state of purity.

---

37 Tosefta Demai ii, 2; cf. Avot de-R. Nathan, Version A, xli, ed. Schechter, p. 132.

38 Thus, for example, Bet Hillel permitted the selling of a draught-cow, used for ploughing, in the Sabbatical Year to a person suspected of not observing the Sabbatical laws (Sheviʻit v, 8).

39 TB Shevuʻot 39a; and cf. TB Sanhedrin 27b.

40 Tosefta Demai ii, 3.

41 He is however permitted to buy dry food from the ʻam ha-aretz: *ibid.*

42 Tosefta Demai ii, 3, 20–22; Avot de-R. Nathan, Version A, xli, ed. Schechter, p. 132. This also follows from Demai ii, 3; vi, 9.

Thus in Mishnah Ḥagigah it is said of Johanan b. Gudgada [a Tanna of the Second Temple period] that היה אוכל על טהרת הקדש כל ימיו "all his life he ate secular food at the degree of purity required for holy things."[43] This was the highest degree of eating secular food in a state of purity, not only in the purity required for secular food but even in that required for holy things.[44]

Of Rabban Gamaliel it was said רבן גמליאל היה אוכל על טהרת חולין כל ימיו והיתה מטפחתו מדרס לקודש "All his days Rabban Gamaliel ate secular food prepared in purity and his apron was a source of midras-impurity to those who ate holy things."[45]

The eating of secular food in a state of purity was in the period of transition to the days of the Amoraim still regarded as an outstanding virtue. Thus it is told of R. Ḥiyya that he instructed his nephew Rav ר׳ חייה רובה מפקד לרב, אין את יכיל מיכול כל שתא חולין בטהרה אכול, ואם לאו, תהא אכיל שבעה יומין מן שתא "R. Ḥiyya the Elder charged Rav, If you can eat secular food in the observance of the rules of ritual purity during the whole year, do so, but if you cannot, eat it seven days in the year."[46]

The aggadah ascribes the eating of secular food in a state of purity

---

43 Ḥagigah ii, 7.

44 See above, pp. 59–60.

45 Tosefta Ḥagigah iii, 2. The reference is apparently to Rabban Gamaliel of Jabneh. The halakhah that follows mentions Onḳelos the proselyte [in Lowe's MS.: Aquila], who was also of the period of Jabneh and who אוכל על טהרת הקודש כל ימיו והיתה מטפחתו מדרס לחטאת "all his days ate secular food at the level of purity required for holy things, and his apron was a source of midras-impurity to those who attended to the water of purification" (Tosefta Ḥagigah iii, 3). Cf. Seder Eliyahu Rabbah xvi, ed. Friedmann, p. 72; and see Lieberman, *Tosefta ki-Peshuṭah*, Seder Mo'ed, pp. 1309–1311.

There is no definite proof that R. Joshua b. Hananiah also ate secular food in a state of purity, as Büchler sought to infer from the fact that R. Joshua, after coming out from discussing some matter with a noblewoman in Rome, took a ritual bath for fear that שמא ניתזה צינורא מפיה "some spittle spurted from her mouth" on to his garments (TB Shabbat 127b). But this is no proof of Büchler's contention, since in the account of the incident it is emphasized that R. Joshua took a ritual bath for the purpose not of eating but of studying the Torah, it being stated that אחר שיצא ירד וטבל ושנה לתלמידיו "after he came out, he went down, had a ritual bath, and taught his disciples," as was indeed the custom, mentioned earlier, of many to study the Torah in a state of purity. See Büchler, *Der galiläische 'Am-ha'Areṣ*, pp. 121–122; and, opposed to him, Allon, *Meḥḳarim*, I, p. 158, note 44.

46 TJ Shabbat i, 3c. This statement reflects the difficulty of eating secular food in a state of purity.

to Abraham[47] and to Saul.[48] This does not of course represent a tradition about the customs of the ancestors of the Jewish people but is rather a statement that reflects the importance assigned in the period under discussion to the eating of secular food in a state of purity.

According to Allon,[49] the fact that certain Sages were singled out for eating secular food in a state of purity does not indicate that it was an additional restriction but rather that it was a commandment obligatory on all Jews. For the special mention made of the Sages who ate secular food in the observance of the rules of purity, Allon gave three reasons. a) In later generations there were not many, even among the Sages, who were heedful of this commandment. b) It was particularly difficult to observe the commandment all the time. c) Alongside the view that the eating of food in a state of purity was obligatory on everyone at all times, there was also the opinion that this obligation was not grounded in the law, as were holy things and terumah.

It would appear that despite Allon's statements — and there is perhaps even support for it in some of his arguments — the eating of secular food in a state of purity was definitely an additional restriction. Not all the Sages, Allon himself admitted, ate secular food in the observance of the rules of purity, and even those who did so, did not do it during the whole year. Some Sages specifically denied any connection between ritual purity and secular food. Even if we agree with Allon's opinion that some Sages regarded the eating of secular food in a state of purity as a commandment binding on everyone, this viewpoint itself may be considered an additional restriction. In summing up it can be said that the eating of secular food in a state of purity was an additional restriction observed in general by individuals and adopted by the ḥaverim who were scrupulous in observing it within the framework of the association.

What follows from all this is the principle that Allon's view, that "the main purpose of the associations was to make it easier for themselves to observe the commandments relating to food,"[50] is not substantiated by the sources and is a distortion of the historical situation.

The mechanism and framework of the association find expression not only in the halakhah pertaining to the conditions of admission, but

---

[47] TB Bava Metziʻa 87a; and cf. Genesis Rabbah xlvii, ed. Theodor-Albeck, pp. 490–491; and parallel passages.

[48] Midrash Psalms, Ps. 7, ed. Buber, p. 63; Numbers Rabbah xi, 3; and parallel passages.

[49] Allon, *Meḥḳarim*, I, pp. 158–169.

[50] *Ibid.*, note 61.

also in related evidences concerning the association's mode of life as a whole.

During his membership in the association the ḥaver had to be scrupulous in fulfilling his obligations as a ḥaver and obedient to the association and its institutions. The expression used by the halakhah in this connection is that the ḥaver had to נענה לחבורה "promise allegiance to the association,"[51] that is, he had to give a report of his actions before the association or its institutions, and was undoubtedly liable to punishment and to the imposition of sanctions if he deviated from his obligations. Matters could reach the stage of דוחין אותו מחבירותו "expelling him from his membership in the association,"[52] at times without any possibility of his reinstatement.

According to R. Meir, the association did not keep a ḥaver who was suspected of not fulfilling even one of its obligations. עם הארץ שקיבל עליו דברי חבירות, ונחשד על דבר אחד נחשד על כולן, דברי רבי מאיר, וחכמים אומרים, אינו חשוד אלא על אותו דבר בלבד "An 'am ha-aretz who took upon himself the obligations of the association and was suspected of one thing, was suspected of all of them. Such is the view of R. Meir. The Sages however said, He is suspected only of that thing."[53] R. Meir also held that whoever was expelled from the association was debarred from returning to it. וכולן שחזרו בהן, אין מקבלין אותן עולמית, דברי רבי מאיר. רבי יהודה אומר, חזרו בהם בפרהסיא מקבלין אותן, במטמוניות אין מקבלין אותן. רבי שמעון ורבי יהושע אומר, לעולם מקבלין אותן, שנאמר ״שובו בנים שובבים״ "And all of them, [even] if they repented, are never readmitted. This is the opinion of R. Meir. R. Judah said, If they repented publicly, they are readmitted, but if secretly, they are not readmitted. R. Simeon and R. Joshua said, They are always readmitted, as it is said, 'Return, O faithless children'."[54] By denying the validity of repentance, R. Meir's approach was far removed from the spirit of the Pharisaic halakhah and applied only to a closed association.

Apparently R. Meir's view here, as in other halakhot relating to an association, reflected the historical situation prevailing in the associations, whereas the statements of the Tannaim who disagreed with him expressed the theoretical halakhic outlook, current in Jewish law in their day, on the laws and customs of an association. Such an approach on

---

51 TJ Demai ii, 23a.

52 Tosefta Demai iii, 4.

53 Tosefta Demai ii, 3; and cf. TJ Demai ii, 22d.

54 Tosefta Demai ii, 9; TB 'Avodah Zarah 7a; TB Bekhorot 31a.

the part of Tannaim we have come across, for example, in R. Judah's statement in the Mishnah Demai that deals with the conditions of admission to an association. It should be pointed out that R. Meir's remarks about a ḥaver who wished to repent contradicted his personal view on the subject of repentance, in that it was he who in point of fact emphasized the power of repentance. So it was in his statements of the type of תניא, היה רבי מאיר אומר, גדולה תשובה שבשביל יחיד שעשה תשובה מוחלין לכל העולם כולו, שנאמר "ארפא משובתם אוהבם נדבה כי שב אפי ממנו," מהם לא נאמר אלא ממנו "It was taught, R. Meir declared, Great is repentance, for on account of an individual who has repented the sins of all the world are forgiven, as it is said, 'I will heal their faithlessness; I will love them freely, for my anger has turned from him.' It does not say, 'from them' but 'from him'."[55] And so it was, too, in R. Meir's conduct. Thus, for example, he was the only Sage who maintained relations with Elisha b. Avuyah even after the latter had become an apostate. This is reflected in particular in the account of Elisha b. Avuyah's last hours, which concluded with a discussion between him and R. Meir on repentance. It is also reflected in R. Meir's remark that דומה שמתוך תשובה נפטר רבי "It seems my teacher died repenting."[56] In general mention should be made of R. Meir's liberal outlook, which also found expression in his attitude to the alien world and in his observations about non-Jewish culture. From all this it is clear that R. Meir's restrictive views of the laws relating to associations did not reflect his personal opinion on the subjects under discussion but represented traditions of his on the actual practice in the associations.

There is not much information on the customs, mode of life, and daily activities of the association. The reason for this is that no work is extant which, compiled by the association, sets forth these details. Our knowledge of the associations is derived from the halakhah, which was redacted and in part also composed after the time of the associations. We are therefore informed of matters which relate to the halakhah, such as, for example, the ways of maintaining purity in the association, and are less informed of the atmosphere and customs which prevailed in it.

Although there is no specific evidence of it in the halakhah, the associations presumably had meals in common.

The term ḥavurah [חבורה], used for a company or an association

---

55 TB Yoma 86b.

56 TJ Ḥagigah ii, 77c; and parallel passages.

although not necessarily that of the ḥaverim, occurs frequently in the halakhah in connection with communal meals. Thus mention is made of חבורת מצוה "the company [celebrating] a religious act" in a reference to religious meals, as, for example, on the occasion of the intercalation of the month.[57] Mention is also made of companies [ḥavurot] for eating the paschal lamb.[58] Much closer to the concept of the ḥavurah of the ḥaverim is the evidence in the Tosefta. אמר רבי אלעזר ברבי צדוק, כך היו חבורות שבירושלם נוהגות, אילו לסעודת אירוסין ואילו לסעודת נישואין "R. Eleazar b. Zadok said, This is what the associations in Jerusalem were accustomed to do, Some [went] to a betrothal feast and some to a wedding feast."[59] An allusion to communal meals may also be found in Mishnah Zavim וכלן טהורין לבני הכנסת וטמאין לתרומה "In both cases however they are pure for members of the synagogue but impure for terumah."[60] Here the expression "members of the synagogue" refers, so it seems, to the ḥaverim who ate their secular food in a state of purity, from which some wish to infer that communal meals were eaten in the synagogue.[61]

One of the central restrictions observed by the association related to food, namely, the eating of secular food in a state of purity. The conditions of admission included stages at which a candidate was accepted for the observance of the rules of purity and for the observance of the purity of liquids. Many of the association's laws referred to the purity of food. All these factors combine to form a background to the view that in the association communal meals were the customary procedure. It should be pointed out that other associations and sects in Israel, such as the Yaḥad and the Essenes, likewise had communal meals, and as in other matters so apparently here, too, there was a similarity between the association of the ḥaverim and these sects. Among the Christian associations of that period there were also the custom of, and a great importance attaching to, communal meals.[62]

---

57 Sanhedrin viii, 2.

58 Pesaḥim vii, 13; viii, 3.

59 Tosefta Megillah iv, 15.

60 Zavim iii, 2.

61 Rabin, *Qumran Studies*, pp. 33–34. In Lowe's MS. and in the Mishnah in the Babylonian Talmud the reading is בית הכנסת "the synagogue" instead of בני הכנסת "members of the synagogue."

62 For the Yaḥad, see, for example, *Manual of Discipline*, vi, 2–8; for the Essenes, see *BJ*, ii, 8, 5, §§129–133; etc. The ḥasidim may also have had communal meals: תרין חסידין הוון באשקלון אכלין כחדא ושתיי כחדא ולעיי באוריתא כחדא "Two ḥasidim [who] were in Ashkelon ate together, and drank together, and studied the

One of the conceptual reasons for eating secular food in a state of purity and for the possible existence of communal meals is to be found in the fact that the meal represented a ceremony corresponding to the offering of a sacrifice, with the table as the altar, and with the blessings before and after the meal corresponding to the prayer before and after the offering of the sacrifice. This view was also prevalent after the destruction of the Second Temple, perhaps then with an even greater intensity, since the table on which the meal was eaten was regarded as a sort of substitute for the ritual of the altar ...דכתיב, ״המזבח עץ שלש. אמות גבוה,״ וכתיב, ״וידבר אלי זה השלחן אשר לפני ה׳,״ פתח במזבח וסיים בשלחן, רבי יוחנן ורבי אלעזר דאמרי תרוייהו, כל זמן שבית המקדש קיים מזבח מכפר על ישראל, ועכשיו שלחנו של אדם מכפר עליו "...As it is written, 'An altar of wood, three cubits high... He said to me, "This is the table which is before the Lord".' The verse begins with 'altar' and ends with 'table.' R. Johanan and R. Eleazar both said, As long as the Temple stood the altar atoned for Israel, but now a man's table atones for him."[63] This may be connected with the halakhah which states that it is not allowed to include an 'am ha-aretz in a zimmun[a)] after a meal ,ותניא אין מזמנין על עם הארץ "And it has been taught, An 'am ha-aretz is not reckoned in for a zimmun."[64]

The association did not exhaust all the possibilities which were available to it within the framework of a closed order and some of which were put into practice in other associations in Israel. Instead it left several areas to the ḥaver's freedom of action and of living.

Torah together" (TJ Ḥagigah ii, 77d; and cf. TJ Sanhedrin vi, 23c). The custom of having common meals is mentioned by Josephus in *AJ*, xiv, 10, 8, §§213–216, in an edict of the days of Julius Caesar. Common meals were also customary among the early Christians, and it is in reference to them that the terms εὐχαριστία, ἀγάπη, which occur in *The Testaments of the Twelve Patriarchs* (διδαχή), are to be explained [Allon, *Meḥkarim*, I, pp. 286–291, has shown the connection between the Christian meals and those of the associations in Israel]. See also *CPJ*, I, p. 254, no. 139, which apparently deals with meals held in common. In Petra special rooms hewn out of the rock and intended for common meals have been found; those participating in them bore a name corresponding to ḥaver. See J. Cantineau, *Le Nabateén*, Paris, 1930–1932, II, p. 63.

63 TB Berakhot 55a; TB Ḥagigah 27a; TB Menaḥot 97a.

64 TB Berakhot 47b.

a) When three or more men partake of a common meal, one of them, who conducts the grace after the meal, commences by inviting the others to join in the recital of the grace, and to each part of the introductory ritual recited by him, they make responses. This ceremony of inviting those present to participate in the saying of the grace after the meal is called zimmun.

This was noticeably so in matters relating to the ḥaver's family in the association, the laws pertaining to which show above all that the association accepted the whole family, and under no circumstances required of a ḥaver that he sever his family ties or even get rid of his slaves in order to be admitted to the association. In these matters the association was generally guided by a preference to maintain the family framework rather than achieve greater scrupulousness in those areas round which the life of the association centred.

The members of the ḥaver's immediate family as well as his slaves had to take upon themselves the obligations of the association, not however in its presence, but in that of the master of the house, the ḥaver. המקבל עליו בפני חבורה, אין בניו ועבדיו צריכין לקבל בפני חבורה, אבל לפניו מקבלין. רבן שמעון בן גמליאל אומר, אינו דומה חבר שקילקל לבן חבר שקילקל "He who takes upon himself [the obligations of a ḥaver] in the presence of an association, his sons and slaves need not take [upon themselves their obligations] before an association but take [them upon themselves] in his presence. Rabban Simeon b. Gamaliel said, A ḥaver who has sinned is not on a par with a ḥaver's son who has sinned."[65] From the disciplinary point of view, too, the members of the family were subject to the master of the house. תני, הוא נענה לחבורה, ובניו ובני ביתו נענין לו "It has been taught, He promises allegiance to the association, and his sons and the members of his family promise allegiance to him."[66] Thus the members of the family were not in fact regarded as ḥaverim with equal rights and obligations in the association, but constituted as it were a periphery of the association that belonged to it passively and whose relation to it was through the ḥaver, the head of the family.

These laws applied also if the immediate or more distant members of the family were 'ammei ha-aretz or the ḥaver brought 'ammei ha-aretz into his family. [This is dealt with in detail below, pp. 161–163.]

In other matters, too, the association was liberal in contrast to the usual procedure in closed associations, and was consequently not cut off from the normative community.

One of the areas in which the association did not follow a different mode of life from the rest of the community was in the economic sphere.

---

65 Tosefta Demai ii, 14. However TB Bekhorot 30b gives a different opinion in the name of Rabban Simeon b. Gamaliel: אף בניו ובני ביתו צריכין לקבל בפני שלשה חבירים "His sons and the members of his family are also required to take [upon themselves their obligations] in the presence of three ḥaverim." See Lieberman, *Tosefta ki-Peshuṭah*, Seder Zera'im, p. 217.

66 TJ Demai ii, 23a.

In the halakhot relating to the association we find no reference to a collective economy or to the abolition of private property. Rather was the opposite the case. From the halakhah it is evident that ḥaverim had their own private property. Thus the halakhah mentions the fields as well as the slaves of ḥaverim, and so on. Although the association as a united body was undoubtedly marked by a large measure of mutual assistance, the maintenance of private property, like the preservation of the family framework, besides engendering no extreme change in the mode of life within the association as compared to that outside it, served to sustain contacts between those who were members of the association and those who were not. In connection with the possession in the associations of private property, mention should be made of the statement שלי שלך ושלך שלי — עם הארץ "He who says, What is mine is yours, and what is yours is mine, is an 'am ha-aretz."[67] This statement sets none other than the 'am ha-aretz, who in contrast to the ḥaver belonged to the lowest social grade, on the same level as members of the sects, such as the Yaḥad, which upheld the principle of the joint ownership of property.[68]

Nor in their places of settlement and of residence did the ḥaverim separate themselves from the normative community, as was the case with other associations among both the Jews and other peoples. No evidences are extant of associations of ḥaverim in the desert or in places remote from the usual settlements. Instead, the opposite is true. The ḥaverim neither left their homes nor concentrated themselves in closed areas in the settlements, but continued to live with the whole community, and even, as we shall see later, in close proximity to and neighbourliness with 'ammei ha-aretz. This, too, necessarily led to intimate, mutual relations between them and the rest of the population.

The association may have expressed its relations with the normative community not only by the non-severance of ties with it, but also by acting as a sort of leaven within the local community in which it was located. If we refer "the associations in Jerusalem," mentioned above, to the association of the ḥaverim, we find in the statement of R. Eleazar b. Zadok, a Tanna at the end of the Second Temple period, instructive evidence of additional customs which prevailed in all or some of the associations. אילו לסעודת אירוסין, ואילו לסעודת נישואין, אילו לשבוע הבן, ואילו

[67] Avot v, 10.

[68] *Manual of Discipline*, i, 11–12; vi, 2–3; etc. See J. Licht, *Megillat ha-Serakhim*, Jerusalem, 1965, pp. 10–13.

ללקט עצמות, אילו לבית המשתה ואילו לבית האבל "Some [went] to a betrothal feast and some to a wedding feast, some to the week of the son and some to the gathering of bones [for burial], some to the house of feasting and some to the house of mourning."[69] Of these customs and commandments which belong to the domain of relief and to the social sphere, the commandments are principally those which require a minyan [a quorum of ten males] and relate to family events. The relief was apparently given by the association not only to its ḥaverim but to the entire population among whom it had its existence.

The customs and commandments mentioned by R. Eleazar b. Zadok as having been observed in "the associations in Jerusalem," and observed, too, in others like them, are attributed in various sources to the ḥever ha-'ir [חבר העיר], the meaning of which is somewhat obscure, but it apparently had some connection with the associations, although not necessarily with all the associations of the ḥaverim nor necessarily in all periods. At all events, we have to reject the explanation of some commentators and scholars who read the expression as ḥaver ha-'ir, meaning a man of distinguished qualities, namely, the Rabbi of the town, and consequently hold that it refers not to an association but to a person. From the context in which the expression occurs the reference is clearly to an association which also acted as a communal body among the population in whose midst it existed, although it is difficult to define precisely the nature of this body.[70]

We find the ḥever ha-'ir actively engaged in rendering assistance in areas relating to mourning and mourners. עולין בחבר עיר על האיש, ואין עולין בחבר עיר על האשה "They go up in a ḥever 'ir [to comfort the

69 Tosefta Megillah iv, 15; Semaḥot xii, 5, ed. Higger, p. 195. See also Mekhilta Aḥariti de-Evel ii, 2–3, ed. Higger, p. 231, where the reading is שבוע הבת ושבוע הבן "the week of the daughter and the week of the son." And see Lieberman, *Tosefta ki-Peshuṭah*, Seder Mo'ed, pp. 1186–1187. On the week of the daughter, see J. Bergmann, "Schebua Ha-ben," *MGWJ*, 76 (1932), pp. 465 ff.

70 On the ḥever ha-'ir, see L. Ginzberg, *Perushim ve-Ḥiddushim ba-Yerushalmi*, New York, 1941–1961, III, pp. 411–432; Lieberman, *Tosefta ki-Peshuṭah*, Seder Zera'im, p. 190; S. Safrai, "Sabbatical Year Commandments under the Conditions Prevailing after the Destruction of the Second Temple" (Hebrew), *Tarbiz*, 35 (1966), pp. 312–314, and the bibliography in note 52; *idem*, "Ha-'Ir ha-Yehudit be-Eretz Yisra'el bi-Teḳufat ha-Mishnah ve-ha-Talmud," *Ha-'Ir ve-ha-Ḳehillah*, Historical Society of Israel, Jerusalem, 1967, pp. 227–236; J. Horovitz, "חבר עיר," *Festschrift zum siebzigsten Geburtstag J. Guttmanns*, Leipzig, 1915, pp. 125–142; Krauss, "חבר עיר," *JJLG*, 17 (1926), pp. 195–241; Horovitz, "Nochmals חבר עיר, Bemerkungen zu des Herrn Prof. Krauss," *ibid.*, pp. 241–315.

mourners if the deceased] is a man, but they do not go up in a ḥever 'ir [to comfort the mourners if the deceased] is a woman.''[71] Of the gathering of bones for burial mention has previously been made. The ḥever ha-'ir also saw to the support of mourners during the period of their mourning.[72]

A ḥever ha-'ir is also referred to in connection with charity. תניא נמי הכי, בני העיר שהלכו לעיר אחרת ופסקו עליהן צדקה נותנין, וכשהן באין מביאין אותה עמהן... במה דברים אמורים, בשאין שם חבר עיר, אבל יש שם חבר עיר תינתן לחבר עיר "It has been taught to the same effect, If the people of one town went to another town and were there rated for a charity contribution, they make the contribution, and on leaving take it with them... When does this rule apply? When there is no ḥever 'ir there, but if there is a ḥever 'ir there, it is given to the ḥever 'ir.''[73] From this halakhah it follows that a ḥever 'ir did not exist in every town, a fact which also removes the concept from the sense of a permanent official institution, and brings it closer to the concept of a voluntary association of ḥaverim.

A ḥever ha-'ir is also mentioned in reference to prayer, which suggests a connection with the associations mentioned in the halakhah concerning a synagogue. רבי אלעזר בן עזריה אומר, אין תפלת המוספין אלא בחבר עיר, וחכמים אומרים, בחבר עיר ושלא בחבר עיר "R. Eleazar b. Azariah said, The musaf [additional] prayers are to be recited only with a ḥever 'ir, but the Sages said, Whether with or without a ḥever 'ir.''[74]

It is difficult to fix the precise limits of the date of the associations, which presumably started among those closely connected with the Temple and with its holy things. This would also explain the numerous halakhot that deal with ḥaverim who were priests.[75]

The associations flourished and developed apparently at the end of the Second Temple period, which was one marked by the decline and weakening of frameworks. Leading as it did to factionalism, this naturally prepared the ground for the formation of closed associations.

Among the halakhot relating to the associations, one deals with חבר ונעשה גבאי "a ḥaver who became a collector,''[76] the reference being

---

71 Semaḥot xi, 2, ed. Higger, p. 187.

72 See also Tosefta Bava Batra vi, 13; TB Ḥullin 94a.

73 TB Megillah 27a–b.

74 Berakhot iv, 7; and see also TB Ro'sh ha-Shanah 34b; TJ Berakhot iv, 8c.

75 This is undoubtedly one of the reasons for Büchler's view that most of the halakhot relating to the ḥaverim deal with priests [see above, p. 5].

76 Tosefta Demai iii, 4; TJ Demai ii, 23a; TB Bekhorot 31a.

presumably to a tax-collector appointed under direct Roman rule. This helps to fix the time when the associations were active, namely, from 6 CE onwards, for it was in this year that direct Roman rule was established.[77]

Mishnah Ḥagigah from i, 9 to the end of the tractate is to all intents and purposes a document which deals with the "purity of abstinence," and which contains rules for those desirous of conducting their lives in the purity of such abstinence. These rules also formed the basis of the associations, some of whose obligations and customs we have traced in our discussion of these Mishnayot. We have previously mentioned that these Mishnayot are a fragment of an early, if not the earliest, tractate, which is to be ascribed to the Second Temple period, and not specifically to the latter part of that period. This, too, provides some support for assigning the associations to the days of the Second Temple.

Since Bet Shammai and Bet Hillel differed on details of halakhot relating to the associations, it follows that the associations existed not only at the time when these disputes took place, but even prior to them, seeing that the differences of opinion between Bet Shammai and Bet Hillel were on matters of detail. This is an additional reason for assigning the associations to the days of the Second Temple.

We have earlier mentioned, and shall yet discuss in greater detail, the considerable similarity that exists between the laws of the association and those of the Essenes and of the Yaḥad, which latter two sects were active in the days of the Second Temple and disappeared after its destruction. In view of this similarity, it may reasonably be assumed that the associations also date from the days of the Second Temple.

The associations apparently continued to exist in the form described above until the destruction of the Second Temple, after which they progressively disappeared. An echo of this is to be found in tractate Soṭah, the concluding Mishnah of which mentions the changes that took place in Israel after the destruction of the Temple. ר׳ פנחס בן יאיר אומר, משחרב

[77] On the structure of Roman rule, see A. Schalit, *Ha-Mishṭar ha-Roma'i be-Eretz Yisra'el*, Jerusalem, 1937. The view of Y. Baer, "The Historical Foundations of the Halakhah" (Hebrew), *Zion*, 17 (1952), pp. 1–55, that the collector is to be identified with the tax-collector in the period of Greek rule is unacceptable. In the halakhot of the association Baer sees reflected the opposition between the farmers who were ḥasidim and the Greek capitalists. E. Schürer, *Geschichte des jüdischen Volkes im Zeitalter Jesu Christi*, Leipzig, 1907[4], I, p. 477, also connects the collector with the farming of taxes prevalent in the days of the Ptolemies. We hear however of no special hatred of tax-collectors in that period.

בית המקדש, בושו חברים ובני חרין וחפו ראשם "R. Phinehas b. Jair said, When the Second Temple was destroyed, ḥaverim and noblemen were ashamed and covered their heads."[78]

Stages in the disintegration of the framework of the associations are reflected in the following Baraita. תנו רבנן, הבא לקבל דברי חבירות צריך לקבל בפני שלשה חבירים, ואפילו תלמיד חכם צריך לקבל בפני שלשה חבירים. זקן ויושב בישיבה אינו צריך לקבל בפני שלשה חבירים, שכבר קיבל עליו משעה שישב. אבא שאול אומר, אף תלמיד חכם אינו צריך לקבל בפני שלשה חבירים, ולא עוד אלא שאחרים מקבלין לפניו "Our Rabbis taught, One who comes to take upon himself the obligations of a ḥaver is required to accept them in the presence of three ḥaverim, and even a talmid ḥakham is required to accept the obligations in the presence of three ḥaverim. An elder, who is a member of the scholars' council, is not required to accept these obligations in the presence of three ḥaverim, having already accepted them when he took his place in the council. Abba Saul said, Even a talmid ḥakham is not required to accept the obligations of a ḥaver in the presence of three ḥaverim, and not only this but others may accept the obligations of a ḥaver in his presence."[79]

The first part of this halakhah reflects the conditions of admission to associations that obtained during the time they existed in the days of the Second Temple. Then life in the associations did not centre round the study of the Torah, nor did talmidei ḥakhamim enjoy any special privileges during the probationary period prior to their admission to the association. The second part of the halakhah, as also the statement of Abba Saul, refer to the time when the associations had ceased to exist, when the title of ḥaver applied to a talmid ḥakham, and when the leaders of the Sages sought to abolish the associations at the fringes of, and the sects outside, the normative community, and to unite all of them around the aim of studying the Torah.

Despite the fact that the associations ceased to exist after the destruction of the Second Temple, the laws governing them did not fall into oblivion. Instead they constituted a subject which extensively occupied the attention especially of the Tannaim of the generation of Usha under

---

78 Soṭah ix, 15. Here בני חרין has the sense of a social aristocracy, from the word חורים, which means men who command respect. Cf. Bava Ḳamma viii, 10. In the Kaufmann MS. the reading is בושו חבירים ובני חבירין "ḥaverim and the sons of ḥaverim were ashamed," which appears to be corrupt (but was preferred by Allon, *Meḥḳarim*, II, pp. 59–60). On this Mishnah as a later addition, see J. N. Epstein, *Mavo le-Nusaḥ ha-Mishnah*, Jerusalem, 1964², II, pp. 949–950, 976–977.

79 TB Bekhorot 30b.

the leadership of R. Meir. Although it represented a discussion of a framework that had already ceased to exist, this exercise did not have merely an academic significance, since it also reflected the desire to make the laws of the association, at least in part, the accepted halakhah for the people as a whole. Thus, for example, on the subject of the eating of secular food in a state of purity תאנא בשם רבי מאיר, כל מי שקבוע בארץ ישראל, ומדבר לשון הקדש, ואוכל פירותיו בטהרה, וקורא קריאת שמע בבקר ובערב — יהא מבושר שבן העולם הבא הוא "It was taught in the name of R. Meir, Whoever is permanently settled in Eretz Israel and speaks Hebrew, eats his produce in a state of purity and recites the shema' morning and evening is assured of a life in the world to come."[80]

These tendencies are to be connected with the general desire, which marked the period following the destruction of the Second Temple, to maintain as far as possible the commandments observed during the existence of the Temple in order thereby to fill the vacuum created by its destruction. The wish moreover to assume additional restrictions, such as the eating of secular food in a state of purity, may be regarded as part of the ascetic trends current among the people after the Great Revolt against the Romans and intensified after the Bar Kokheva revolt in a fresh wave of asceticism during the generation of Usha.[81]

That there was no desire to renew the framework of the associations in the days of Usha can be inferred from the halakhot, for they deal generally with the individual ḥaver and with the areas in which he was obliged to be scrupulous, but not with the community of the ḥaverim, that is, with the association as a whole.

As we have seen, R. Meir often quoted traditions from halakhot relating to ḥaverim that were current in the associations in the Second Temple period. R. Meir not only adopted the stricter view in any dispute about these halakhot[82] but also, in both his personality and outlook, lived up to the high standards he proclaimed. His advocacy of abstinence and purity led to his being given the title of "ḳadosh," saint, a title granted to but few Sages. R. Jose b. Ḥalafta, his contemporary and intimate friend, called him אדם גדול, אדם קדוש, אדם צנוע "A

[80] TJ Sheḳalim iii, 47c.

[81] See E. E. Urbach, "Ascesis ve-Yissurim be-Torat Ḥazal," *Sefer ha-Yovel li-Kevod Yitzḥaḳ Baer*, Jerusalem, 1960, pp. 48–68; *idem*, *Ḥazal — Pirḳei Emunot ve-De'ot*, Jerusalem, 1969, pp. 392–396.

[82] See, for example, Tosefta 'Avodah Zarah iii, 11, and parallel passages; Tosefta Demai ii, 1, 3, 9, 17, and parallel passages; etc.; etc.

great man, a saintly man, a virtuous man.''[83] R. Meir himself frequently used the concept of holiness which is closely connected with the concepts of abstinence and purity.[84]

Some of R. Meir's pupils formed themselves into an association known as קהילא קדישא דבירושלים "the Holy Congregation in Jerusalem," in which they adopted patterns of living in part resembling those of the associations in the days of the Second Temple. Describing the mode of life that prevailed in this association, the Midrash states ...ולמה הוא קורא אותן עדה קדושה, ששם היו רבי יוסי בן משולם, ורבי שמעון בן מנסיא, שהיו משלשין היום — שליש לתורה, שליש לתפילה, שליש למלאכה, ויש אומרים, שהיו יגעין בתורה בימות החורף, ובמלאכה בימות הקיץ "And why does he call them a holy community? Because it included R. Jose b. Meshullam and R. Simeon b. Menasya who divided the day into three, one-third devoted to the Torah, one-third to prayer, and one-third to work. Some say they studied the Torah during the winter days and worked during the summer days.''[85] It may be assumed with a great deal of probability that in addition to being occupied in the study of the Torah, in prayer, and in work, the members of this association, like those of the associations in the days of the Second Temple, were also scrupulous in observing purity. The very term "holy" by which the association was known is, as we have seen, close to the concept of purity, as, for example, in the statement of R. Phinehas b. Jair, who lived in that period טהרה מביאה לידי פרישות, ופרישות מביאה לידי קדשה "Purity leads to abstinence, abstinence leads to holiness.''[86] We also find members of the Holy Congregation in Jerusalem adopting the stricter view in halakhot relating to a ḥaver and to purity. העיד רבי יוסי בן המשולם משום רבי נתן אחיו, שאמר משום רבי אלעזר חסמא, שאין עושין טהרות לעם הארץ וכו׳ "R. Jose b. Hameshullam testified in the name of R. Nathan, his brother, who said in the name of R. Eleazar Ḥisma, Food may not be prepared in ritual purity for an ʻam ha-aretz...''[87] In the case of the Holy Congregation in Jerusalem life within the framework of the association was also presumably marked by common meals.[88]

---

[83] TJ Berakhot ii, 5b; TJ Moʻed Ḳaṭan iv, 82d. Resh Laḳish called R. Meir פה קדוש "the saintly mouth" (TB Sanhedrin 83a).

[84] TB Berakhot 17a.

[85] Ecclesiastes Rabbah ix, 9.

[86] Soṭah ix, 15; TJ Shabbat i, 3b.

[87] Tosefta Demai iii, 1.

[88] For ḥasidim and pupils, who studied the Torah together and had meals in common, see TJ Ḥagigah ii, 77d; TJ Sanhedrin vi, 22c.

Like the associations of Second Temple times the Holy Congregation in Jerusalem, too, did not separate itself from the normative community. Together with other Sages of that period, they participated in the formulation and crystallization of the halakhah. In several passages in the Babylonian Talmud halakhic and aggadic statements are quoted in the name of the Holy Congregation in Jerusalem.[89] R. Judah ha-Nasi and R. Joshua b. Levi transmitted halakhot in their name, and the heads of the Holy Congregation in Jerusalem are mentioned in the Mishnah.[90] Mishnayot of theirs are also quoted anonymously.[91] [92]

Several scholars of the Dead Sea sect as well as others[93] have sought to compare, and at times even to identify, the association with the community of the Yaḥad or with the Essenes. Some scholars have even compared the association with the schools of the Pythagoreans.

It is in particular customary to compare the association both with the Essenes as described by Josephus and with the community of the Yaḥad and its laws as contained in the *Manual of Discipline*. A most notable similarity is to be found in their respective laws of admission.

The laws of admission to the sect of the Essenes, as given by Josephus, are as follows.

Τοῖς δὲ ζηλοῦσιν τὴν αἵρεσιν αὐτῶν οὐκ εὐθὺς ἡ πάροδος, ἀλλ' ἐπὶ ἐνιαυτὸν ἔξω μένοντι τὴν αὐτὴν ὑποτίθενται δίαιταν, ἀξινάριόν τε καὶ τὸ προειρημένον περίζωμα καὶ λευκὴν ἐσθῆτα δόντες. ἐπειδὰν δὲ τούτῳ τῷ χρόνῳ πεῖραν ἐγκρατείας δῷ, πρόσεισιν μὲν ἔγγιον τῇ διαίτῃ καὶ καθαρωτέρων τῶν πρὸς ἁγνείαν ὑδάτων μεταλαμβάνει, παραλαμβάνεται δὲ εἰς τὰς συμβιώσεις οὐδέπω. μετὰ γὰρ τὴν τῆς καρτερίας ἐπίδειξιν δυσὶν ἄλλοις ἔτεσιν τὸ ἦθος δοκιμάζεται καὶ

---

[89] TB Berakhot 9b; TB Betzah 14b; etc.

[90] Ḥagigah i, 7; Terumot iv, 7; etc.

[91] Cf. Tosefta Bekhorot iv, 11 with Mishnah Bekhorot vi, 8; etc.

[92] On the Holy Congregation in Jerusalem and for other views on it, see S. Safrai, "The Holy Congregation in Jerusalem" (Hebrew), *Zion*, 22 (1957), pp. 183–193; *idem*, "The Holy Congregation in Jerusalem," *Scripta Hierosolymitana*, XXIII, Studies in History, 1972, pp. 62–78, and see the bibliography *ad loc.*

[93] Licht, *Megillat ha-Serakhim*; H. Mantel, "The Nature of the Great Synagogue (Knesset ha-Gedolah)," *Fourth World Congress of Jewish Studies*, I, Jerusalem, 1967, pp. 81–88 (Hebrew), p. 258 (English Summary); Ch. Rabin, *Ha-Yaḥad, ha-Ḥavurah ve-ha-Issiyyim*, *'Iyyunim bi-Megillot Midbar Yehudah*, Jerusalem, 1957, pp. 104–122; and see Urbach, *Ḥazal*, pp. 521–525; Lieberman, *JBL*, 71 (1952), pp. 199–206; Rabin, *Qumran Studies*; Neusner, *HTR*, 53 (1960), pp. 125–142. On similar associations among non-Jews, see I. Lévy, *La légende de Pythagore de Grèce en Palestine*, Paris, 1927, pp. 236–263.

φανεὶς ἄξιος οὕτως εἰς τὸν ὅμιλον ἐγκρίνεται. πρὶν δὲ τῆς κοινῆς ἅψασθαι τροφῆς ὅρκους αὐτοῖς ὄμνυσι φρικώδεις, πρῶτον μὲν εὐσεβήσειν τὸ θεῖον, ἔπειτα τὰ πρὸς ἀνθρώπους δίκαια φυλάξειν κτλ.

"A candidate anxious to join their sect is not immediately admitted. For one year, during which he remains outside the fraternity, they prescribe for him their own rule of life, presenting him with a small hatchet, the loin-cloth already mentioned, and white raiment. Having given proof of his temperance during this probationary period, he is brought into closer touch with the rule and is allowed to share the purer kind of holy water, but is not yet received into the meetings of the community. For after this exhibition of endurance, his character is tested for two years more, and only then, if found worthy, is he enrolled in the society. But, before he may touch the common food, he is made to swear tremendous oaths: first that he will practise piety towards the Deity, next that he will observe justice towards men..."[94]

The following are the laws governing admission to the community of the Yaḥad, as set out in the *Manual of Discipline*.

וכול המתנדב מישראל להוסיף על עצת היחד ודורשהו האיש הפקיד ברואש הרבים לשכלו ולמעשיו. ואם ישיג מוסר יביאהו בברית לשוב לאמת ולסור מכל עול. והבינהו בכול משפטי היחד.

ואחר בבואו לעמוד לפני הרבים ונשאלו הכול על דבריו. וכאשר יצא הגורל על עצת הרבים יקרב או ירחק. ובקורבו לעצת היחד לוא יגע בטהרת הרבים עד אשר ידרושהו לרוחו ומעשו עד מולאת לו שנה תמימה. וגם הואה אל יתערב בהון הרבים.

ובמולאת לו שנה בתוך היחד ישאלו הרבים על דבריו לפי שכלו ומעשיו בתורה. ואם יצא לו הגורל לקרוב לסוד היחד על פי הכוהנים ורוב אנשי בריתם יקרבו גם את הונו ואת מלאכתו אל יד האיש המבקר על מלאכת הרבים. וכתבו בחשבון בידו ועל הרבים לוא יוציאנו. אל יגע במשקה הרבים עד מולאת לו שנה שנית בתוך אנשי היחד.

ובמולאת לו השנה השנית יפקודהו על פי הרבים. ואם יצא לו הגורל לקרבו ליחד יכתובהו בסרך תכונו בתוך אחיו לתורה ולמשפט ולטוהרה ולערב את הונו ויהי עצתו ליחד ומשפטו.[95]

"Anyone of Israel who of his own free will desires to join the community of the Yaḥad,[a] the superintendent at the head of the Rabbim[b]

[94] *BJ*, ii, 8, 7, §§137–142.

[95] *Megillat ha-Serakhim*, vi, 14–23, ed. Licht, pp. 149–151.

a) Lit. "together." The term apparently applied to the entire sect, that is, to both full members and candidates.

b) Lit. "the many." These were apparently the full members of the sect. And see Licht, *Megillat ha-Serakhim*, p. 117.

shall examine him with regard to his understanding and his actions, and if he is found to be amenable to discipline, he shall admit him to the covenant whereby he undertakes to return to truth and to depart from all unrighteousness. And he shall acquaint him with all the laws of the Yaḥad.

Afterwards when he comes and appears before the Rabbim, all of them shall be asked concerning him. According to the result of the ballot following the deliberation of the Rabbim, he shall be admitted or rejected. If he is admitted to the community of the Yaḥad, he shall not touch the food of the Rabbim that has been prepared in purity, until at the end of a whole year they shall have examined him concerning his spirit and his actions. Nor shall his property be pooled with that of the Rabbim.

After he shall have been a full year in the Yaḥad, the Rabbim shall inquire about him as to his understanding and his actions with regard to the Torah. And if the ballot in accordance with the view of the priests and the majority of the members of their covenant be to admit him to the followship of the Yaḥad, his property and his earnings from his labours shall be given to the official in charge of the work of the Rabbim. And they shall be entered in an account kept by him and shall not be spent on the Rabbim. He shall not touch the liquids of the Rabbim until he shall have been two years among the members of the Yaḥad.

At the end of the second year the Rabbim shall examine him, and if the ballot be to admit him to the Yaḥad, he shall be enrolled according to his rank among his brethren, [as a full member with them] in respect of the Torah, of judgment, and of purity, and in respect also of the pooling of his property, and he shall participate with his counsel and his judgment in the Yaḥad."

While there are points of similarity between the ḥaverim and the Essenes, any comparison between them is fraught with danger because of the sparse, fragmentary, and vague nature of the information at our disposal about the Essenes. Yet even from the available information it can readily be seen that the association of the ḥaverim and the sect of the Essenes differed greatly from each other.

Whereas purity was indeed central alike to the ḥaverim and to the Essenes, among both of whom it was included in the process of admission, in the association of the ḥaverim there was no hint of the other stages of admission found among the Essenes. These included endurance tests for living in the desert under conditions of hardship, character tests, and an oath, which occupied a much more central place among them than it did among the ḥaverim, if the latter had one at all. This is symptomatic. For, unlike the association, the organization of the Essenes was a means of promoting a secret doctrine, and their very framework a way of safeguarding the mysteries of that doctrine and the spiritual and physical purity of those privileged to be the vessels for its reception.

A much greater similarity existed between the association of the ḥaverim and the Yaḥad. From a comparison of the stages of admission to them it is clear that in both there were a public undertaking to accept their respective obligations, a period devoted to learning the laws, and periods of trial and of gradual admission defined and conditioned by the extent of the candidate's trustworthiness in matters concerned with purity. In both, too, the last stage of admission related to the purity of liquids.[96]

And yet, despite all these points of similarity, it cannot be maintained that the reference is to the same, or even to one type of, organization.

Although the conditions of admission appear at first sight to be so similar, they are not necessarily alike. It is sufficient to recall that several terms which occur in the conditions of admission of the ḥaverim, such as kenafayim, as well as the composition of the process of admission as a whole, are not altogether clear. In any event it is still more complicated, if not impossible, to compare the two processes and stages of admission in their entirety.

In their very nature these sects exhibited a basic difference, one whose central point in the case of the Yaḥad, but not of the ḥaverim, was its severance from the normative community. In the Yaḥad we do not hear of a member's continuing relations with his family, nor is there any evidence of the presence of women in the sect. As against this, the association preserved the family framework even if all its members did not join the association. In the various stages of admission to the Yaḥad there was a gradual pooling of the candidate's property with that of the community, in contrast to the situation that obtained among the ḥaverim, among whom even a ḥaver with full membership in the association still continued to own private property. Nor do we have any evidence of the association's having property in common. The ḥaverim

---

96 For the conditions of admission embodied in a unified system, see Lieberman, *JBL*, 71 (1952), pp. 199–206.

did not even shun the 'am ha-aretz completely, despite the latter's disregard of all that was sacred to them. In certain instances they were prepared to recognize a partial obligation on the part of a candidate wishing to join the association, something that was quite inconceivable in the case of the Yaḥad.

In general it may be said that, from the viewpoint of the social classes and the prevailing climate in the nation at the end of the Second Temple period, the ḥaverim were closer to the 'ammei ha-aretz than to the sects of the Essenes and of the Yaḥad, for these latter two referred to whomsoever did not join them by such expressions as אנשי חמס "men of violence," אנשי שחת "men of the pit" [that is, eternally damned], and other similar phrases, evidences of disparagement and of social avoidance. Yet in their motivation and standard of conduct there was naturally a greater affinity between the ḥaverim and the Essenes and the Yaḥad than between the ḥaverim and the 'ammei ha-aretz.

To sum up: the similarities between the ḥaverim and the Essenes and the Yaḥad, while reciprocally instructive regarding the nature of these sects, cannot be used as evidence of their identity. Yet among the things which can be reciprocally deduced from them is, for example, the date of the association of the ḥaverim, which presumably corresponded to that of the Essenes and the Yaḥad, that is, the end of the Second Temple period. But, as previously stated, we cannot go much beyond this and establish the identity of these sects. Eretz Israel in the first century CE teemed with various sects and associations, so that it is not surprising that there should at that time have been kindred groups, such as the Essenes, the Yaḥad, and the ḥaverim. But it would be wrong, by doing violence to texts or disregarding details, to identify them with one another.

Another social concept, which relates to both the ḥaverim and the 'ammei ha-aretz as well as to the laws of the tithes and the extent to which these laws were observed, is that of the ne'eman, the trustworthy person.

In the Mishnah it is said of the ne'eman המקבל עליו להיות נאמן, מעשר את שהוא אוכל ואת שהוא מוכר ואת שהוא לוקח, ואינו מתארח אצל עם הארץ. רבי יהודה אומר, אף המתארח אצל עם הארץ נאמן. אמרו לו, על עצמו אינו נאמן, כיצד יהא נאמן על של אחרים "If a person takes upon himself to be a ne'eman, he must tithe whatever he eats and whatever he sells and whatever he buys, and he may not be the guest of an 'am ha-aretz. R. Judah said, One who is the guest of an 'am ha-aretz may still be considered a ne'eman. But they said to him, If he is not a ne'eman in respect of himself, how can he be considered a ne'eman in respect of others?"[97]

According to this source, a ne'eman was one who was so scrupulous in observing the laws relating to the tithes that he tithed all produce coming into his possession. This obligation did not, of course, apply to produce which had definitely been tithed, but in every other instance he had to separate the tithes and not rely on either the seller or the buyer.

The question that arises is whether the concept of the ne'eman referred to a social stratum, similar to the concepts of an 'am ha-aretz and a ḥaver, or whether it indicated anyone who was particular only about separating tithes.

Several scholars[98] see in the ne'eman a concept which reflected a social stratum and embodied more than merely a scrupulous attention to the separation of tithes. According to them, the ne'eman is to be regarded as a candidate who, aspiring to the degree of a ḥaver, was at the first of the various stages of admission to the association of ḥaverim.

What prompted this view is that various sources include, in halakhot dealing with ḥaverim and with the association, laws which refer to the ne'eman. Furthermore, the obligations imposed upon the ne'eman concern an area about which the association was particularly scrupulous, namely, the proper separation of tithes.

These scholars seek to substantiate their view by comparing the laws and customs of the Yaḥad as reflected in the Dead Sea Scrolls with the halakhot relating to, and the customs practised by, the ḥaverim and the association. This general comparison reveals, so they maintain, a similarity between the status of a candidate for membership in the Yaḥad and that of a ne'eman.[99]

---

[97] Demai ii, 2; and cf. Tosefta Demai ii, 2.

[98] See Rabin, *Qumran Studies*; Neusner, *HTR*, 53 (1960), pp. 125–142; and cf. Lieberman, *JBL*, 71 (1952), pp. 199–206.

[99] The acceptance of a candidate for membership in the Yaḥad is given in the description of the probationary period. וכול המתנדב מישראל להוסיף על עצת היחד ודורשהו האיש הפקיד ברואש הרבים לשכלו ולמעשיו. ואם ישיג מוסר יביאהו בברית לשוב לאמת ולסור מכול עול. והבינהו בכול משפטי היחד "Anyone of Israel who of his own free will desires to join the community of the Yaḥad, the superintendent at the head of the Rabbim shall examine him with regard to his understanding and his actions, and if he is found to be amenable to discipline, he shall admit him to the covenant whereby he undertakes to return to truth and to depart from all unrighteousness. And he shall acquaint him with all the laws of the Yaḥad" (*Manual of Discipline*, vi, 13–15). The *Damascus Document*, xv, 7–10, gives a different account. ביום דברו עם המבקר אשר לרבים יפקדוהו בשבועת הברית אשר כרת משה עם ישראל, את

Further proof that the ne'eman was a candidate who desired to join the association, comparable to a candidate for the Yaḥad, is, they hold, to be found in a comparison between the penalties inflicted in these two sects. A member of the Dead Sea sect who committed an offence was punished by being excluded from food prepared in ritual purity,[100] and was accordingly in the position of one "admitted to the covenant," that is, he reverted to the status of a candidate for membership in the sect who had obligations imposed upon him but was not yet permitted to touch the food of the Rabbim prepared in ritual purity. A similar penalty was inflicted on a ḥaver who had deviated from the right path. **בראשונה היו אומרין, חבר ונעשה גבאי דוחין אותו מחבירותו, חזרו לומר, כל זמן שהוא גבאי אינו נאמן, פירש מגבייתו הרי זה נאמן** "At first they said, A ḥaver who becomes a tax-collector is excluded from membership of the association. But subsequently they said, So long as he is a tax-collector he is not a ne'eman, but if he gives up his position as a tax-collector he is a ne'eman."[101] This means that a ḥaver who transgressed reverted to the status of a candidate, that is, of being a ne'eman.

Without going into all the doubts raised by comparing the association of ḥaverim with the Yaḥad and referred to above, we believe that the comparison with being "admitted to the covenant," mentioned in the Dead Sea Scrolls, is forced. Greater than any similarity between a ne'eman and a candidate desirous of joining the Yaḥad is the difference between them. Whereas in the Yaḥad anyone wishing to be "admitted to the covenant" proclaimed it only before the **האיש הפקיד** "the superintendent," someone wishing to be a ne'eman had to take it upon himself before the entire association.[102] Nor is there any similarity between the obligations of a ne'eman, which were expressed by a scrupulous observance of the tithes, and those of the candidate admitted to the covenant which, according to the *Manual of Discipline*, took the form of an undertaking **לשוב לאמת** "to return to truth" and **לסור מכל עול** "to depart from all unrighteousness," and, according to

---

הברית לשוב אל תורת משה בכל לב ובכל נפש אל הנמצא לעשות בם "On the day that he speaks with the overseer of the Rabbim, he shall examine him with the oath of the covenant which Moses made with Israel, the covenant to return to the Torah of Moses with a full heart and a full soul to whatever can be done with them."

100 *Damascus Document*, ix, 21; and, similarly, the *Manual of Discipline*, vii, 3; etc.

101 Tosefta Demai iii, 4; TJ Demai ii, 23a; TB Bekhorot 31a.

102 **אבל ברבים אינו נאמן, עד שיקבל עליו ברבים** "But among the Rabbim he is not a ne'eman until he takes it upon himself before the Rabbim" (TJ Demai ii, 22d).

the *Damascus Document*, to observe the commandments of the Torah.

In the punishment, too, whereby the one committing an offence reverted to the stage of being a candidate, there is much more that is obscure than explicit. It cannot be asserted that מובדל מן הטהרה being "separated from food prepared in purity" is identical with בא בברית being "admitted to the covenant," and that we have here the renewed candidature of one who made every effort to achieve a complete return to the sect. True, both did not touch food prepared in purity, but it may be that the one did not do so because he was still a candidate, while the other was unable to do so because of the penalty imposed upon him. Also from the halakhah that deals with a ḥaver who became a tax-collector it is difficult to conclude that his being declared a ne'eman once he has repented is intended to place him in the category of a candidate for membership in the association. It must be stressed that the version of this Baraita differs in the parallel passages: the Jerusalem Talmud has יצא מגבייתו הרי הוא כחבר "If he resigned his office as a tax-collector he is like a ḥaver";[103] while the Babylonian Talmud in Bekhorot has פירש הרי הוא ככל אדם "If he gave it up he is like anybody else."[104] In these parallel passages no mention whatsoever is made of a ne'eman. Moreover, the halakhah itself declares that the qualified exclusion of a ḥaver who became a tax-collector was a new regulation. It is very likely that the statement in the halakhah which follows "At first" reflects the situation that actually obtained in the associations, while the modification of this halakhah that such a ḥaver was not permanently excluded is later and reflects the halakhah at the time it was formulated and not the procedure which in fact prevailed in the associations.

It should be pointed out that in the halakhot dealing with the obligations which the ne'eman took upon himself no mention is made of his being accepted into any society whatsoever, in contrast to the ḥaver, of whom it is distinctly stated that he was admitted to the association. On the other hand, whatever was imposed upon the candidate for membership in the Yaḥad was, so it is specifically mentioned, within the context of his candidature, the final object of which was admission to the covenant of the sect.

In general it may be emphasized that the term ne'eman in the halakhah does not refer to trustworthiness in respect specifically of the

---

103 TJ Demai ii, 23a.

104 TB Bekhorot 31a.

tithes, for it occurs also in other senses. The feature common to these concepts is that the ne'eman is contrasted with one suspected of not observing some commandment or other. For example, הוא [ר׳ ראובן בן אצטרובלי] היה אומר, ג׳ נאמנים הן. עני שהוא נאמן על הפקדון אין נאמן גדול מזה. בעל הבית שהוא נאמן על מעשרותיו אין נאמן גדול מזה. רווק שגדל במדינה ולא נחשד על הערוה אין נאמן גדול מזה "He [R. Reuben b. Estrobile (Strobilus)] said, There are three who are ne'emanim. A poor man who is trustworthy with regard to a deposit: there is no greater ne'eman than he. A householder who is trustworthy with regard to his tithes: there is no greater ne'eman than he. A bachelor who grew up in a large city and was not suspected of licentiousness: there is no greater ne'eman than he."[105] In certain matters even an 'am ha-aretz is referred to as a ne'eman. For example, ועל כלם [על ביצים, דגים וציר], עם הארץ נאמן לומר, טהורין הם, חוץ משל דגה "Concerning all these [eggs, fish, and brine], an 'am ha-aretz is regarded as a ne'eman if he declares them to be pure, except [in the case of the brine] of fish,"[106] or והתניא, נאמן עם הארץ לומר פירות לא הוכשרו, אבל אינו נאמן לומר פירות הוכשרו אבל לא נטמאו "For it was taught, An 'am ha-aretz is regarded as a ne'eman if he says, The produce has not been rendered susceptible to impurity, but he is not regarded as a ne'eman if he says, The produce has been rendered susceptible to impurity but has not been made impure."[107] [108]

An analysis of the halakhah itself thus shows that there is little foundation for the suggestion that the ne'eman was a candidate for admission to the association. A different theory on the nature of the concept must accordingly be put forward.

The connection of the ne'eman with the association derived not from his candidature but from the relation between them, a relation expressed by the ne'eman's taking upon himself, in the presence of the association, the observance in all their details of the laws of the tithes, and hence the ḥaver could have unrestricted economic contact with the ne'eman in respect of tithes and could trade with him without any misgiving. In the halakhot the ne'eman is contrasted with those suspected of not observing the laws relating to tithes and especially with the 'ammei ha-aretz. Hence many halakhot testify to the possibility of main-

---

105 Avot de-R. Nathan, Version B, xxxv, ed. Schechter, p. 78; and cf. TB Pesaḥim 113a.

106 Makhshirin vi, 3.

107 TB Ḥagigah 22b; and see TJ Demai ii, 22d.

108 In general an affinity may be seen between the concept of the ne'eman and the term πιστός in the New Testament.

taining free economic contact with the ne'eman without fear of ṭevel, but not with an 'am ha-aretz or with one who was not a ne'eman in respect of tithes, with whom such contact was prohibited.

That no impure produce might be passed on to a ḥaver, the ne'eman's observance of the laws of the tithes was presumably combined with a certain scrupulousness in matters relating to purity.[109]

The halakhah enjoins a strictness not only in the purchase but also in the sale of produce. The following Mishnah, for example, testifies to the sale of produce specifically to a ne'eman in respect of tithes. לא ימכר אדם את פרותיו משבאו לעונת המעשרות למי שאינו נאמן על המעשרות "No person may sell his produce after the season for tithing has arrived to one who is not a ne'eman in respect of tithes."[110] We have previously mentioned halakhot which, similar to this one, prohibit a ḥaver from transferring food into the possession of an 'am ha-aretz in case he fails to tithe it or renders it impure.

3. The Relations between the 'Ammei ha-Aretz and the Pharisees

Like the ḥaverim, the Pharisees, too, shunned the 'am ha-aretz, mainly for fear of the impurity attaching to him. Evidence of this occurs in an ancient Mishnah which we have often quoted. בגדי עם הארץ מדרס לפרושין "The garments of an 'am ha-aretz are a source of midras-impurity to Pharisees."[111] Here there may be reflected the situation which preceded the formation of the associations of the ḥaverim.

The care to be exercised by Pharisees about coming in contact with the 'ammei ha-aretz is mentioned in disputes between Bet Shammai and Bet Hillel. For example, בית שמאי אומרים, לא יאכל זב פרוש עם זב עם הארץ, ובית הילל מתירין "Bet Shammai said, A Pharisee who has a discharge shall not eat with an 'am ha-aretz who has a discharge, but Bet Hillel permit it."[112] A tradition in the Jerusalem Talmud gives the reasons of Bet Shammai and Bet Hillel in this dispute. ומה טעמיהון דבית הלל, זה זב וזה זב; ומה טעמהון דבית שמאי, שהוא מתרגל עמו בימי טומאתו, הוא

---

[109] We do not deny the possibility of a probationary stage for a candidate for membership in an association, but rather the identification of a ne'eman with such a candidate.

[110] Ma'aserot v, 3.

[111] Ḥagigah ii, 7.

[112] Tosefta Shabbat i, 15.

מתרגל עמו בימי טהרתו "And what is Bet Hillel's reason? The one has a discharge and the other has a discharge. And what is Bet Shammai's reason? If he frequents his company when he himself is impure, he will frequent his company when he himself is pure."[113] Thus both Bet Shammai and Bet Hillel enjoined caution against the impurity of the ʻam ha-aretz, except that Bet Shammai, whose reason is very probably that given in the tradition in the Jerusalem Talmud, adopted the strict view also in the case in which the Pharisee had a discharge, when his impurity was naturally augmented through contact with the ʻam ha-aretz.

Generally it can be said that Bet Shammai adopted a stricter approach than did Bet Hillel in disputes concerning a Pharisee and an ʻam ha-aretz or a ḥaver and an ʻam ha-aretz. Thus, as a further example, בית שמאי אומרים, לא ימכור אדם את זיתיו אלא לחבר, בית הלל אומרים, אף למעשר "Bet Shammai said, A man shall not sell his olives except to a ḥaver, but Bet Hillel said, He may do so also to one who separates tithes."[114] The existing halakhot in this sphere are too sparse to draw far-reaching conclusions from them, such as that the world of the ḥaverim was closer to Bet Shammai than to Bet Hillel. It is very likely that, in keeping with their generally strict approach in the halakhah as opposed to the principle adopted by Bet Hillel, Bet Shammai were stricter in halakhot which required caution on the part of the Pharisees and of the ḥaverim in their contacts with the ʻammei ha-aretz.

Despite such caution exercised by the Pharisees, this did not lead to an estrangement between them and the ʻammei ha-aretz.

Of the incidents which clearly prove this there was the one connected with the ceremony of the willow-branch on the Sabbath during the festival of Tabernacles, when the ʻammei ha-aretz supported the Pharisees.

The nature of the commandment of the willow-branch is set out in the Mishnah. מצות ערבה כיצד, מקום היה למטה מירושלם ונקרא מוצא. יורדין לשם ומלקטין משם מרביות של ערבה, ובאין וזוקפין אותן בצדי המזבח, וראשיהן כפופין על גבי המזבח. תקעו והריעו ותקעו. בכל יום מקיפין את המזבח פעם אחת, ואומרים, ״אנא ה׳ הושיעה נא, אנא ה׳ הצליחה נא״... ואותו היום מקיפין את המזבח שבעה פעמים. בשעת פטירתן מה הן אומרים, ״יפי לך מזבח, יפי לך מזבח,״ רבי אליעזר אומר, ״ליה ולך מזבח, ליה ולך מזבח״ "How was the commandment of the willow-branch [carried out]? There was a place below Jerusalem called Motza. They went down there, gathered from it young willow-

---

[113] TJ Shabbat i, 3c.

[114] Demai vi, 6; and see Tosefta Maʻaserot iii, 13.

branches, and then came and fixed them on the sides of the altar so that their tops bent over the altar. They then sounded a teḳi'ah [a long blast], a teru'ah [a tremulous blast], and again a teḳi'ah. Every day they went round the altar once, saying, 'Save us, we beseech thee, O Lord! O Lord, we beseech thee, give us success!'... But on that day [the seventh day of the festival of Tabernacles] they went round the altar seven times. When they departed, what did they say? 'Yours, O altar, is the beauty! Yours, O altar, is the beauty!' R. Eliezer declared, [They said,] 'To the Lord and to you, O altar. To the Lord and to you, O altar'."[115]

The Pharisaic halakhah laid down that as a rule the commandment of the willow-branch did not supersede the Sabbath except when the seventh day of Tabernacles fell on a Sabbath, in which case the commandment of the willow-branch overrode the Sabbath, and they went round the altar seven times.[116] On this point the Sadducees disagreed with the Pharisees and held that under no circumstances did the commandment of the willow-branch override the Sabbath, even on the seventh day of Tabernacles.[117]

This dispute found practical expression in an incident which is referred to in the Tosefta and given in detail in a Baraita in TB Sukkah. **איתיביה, לולב דוחה את השבת בתחלתו וערבה בסופו — פעם אחת חל שביעי של ערבה להיות בשבת, והביאו מרביות של ערבה מערב שבת, והניחום בעזרה, והכירו בהן בייתוסין ונטלום וכבשום תחת אבנים, למחר הכירו בהן עמי הארץ, ושמטום מתחת האבנים, והביאום הכהנים וזקפום בצידי המזבח, לפי שאין בייתוסין מודים שחיבוט ערבה דוחה את השבת** "He raised an objection against him. The ceremony of the lulav overrides the Sabbath on the first day and that of the willow-branch on the last day [of the festival of Tabernacles]. Once the seventh day of [the ceremonial of] the willow-branch fell on a Sabbath and they brought young willow-branches on the Sabbath eve and placed them in the Temple courtyard. The Boethusians, having discovered them, took and hid them under some stones. On the next

---

[115] Sukkah iv, 5.

[116] Sukkah iv, 3. According to the non-talmudic [ancient ?] halakhah in the Book of Jubilees (16:31), the altar was to be circled seven times every day. In the halakhah there may thus already be a retrogression as compared to the ancient custom.

[117] Basically the Sadducees may not have admitted the validity of the custom of the willow-branch even on weekdays, since it is not mentioned in the Bible, but their opposition increased and assumed a practical form when it came to using the willow-branch on the Sabbath. Cf. S. Safrai, *Ha-'Aliyah le-Regel bi-Yemei Bayit Sheni*, Tel Aviv, 1965, p. 192, and the notes on p. 212.

day some of the ʿammei ha-aretz discovered and removed them from under the stones, and the priests brought them in and fixed them on the sides of the altar. The reason [for hiding the willows] was that the Boethusians did not agree that the beating of the willow-branch overrides the Sabbath."[118] The willow-branches were fixed on the sides of the altar by the priests,[119] and this naturally intensified the opposition of the Sadducees and the Boethusians[120] to the Pharisaic halakhah on the subject, for dominating the priesthood as they did, they sought to prevent the carrying out of something which was contrary to their view. In this incident the willow-branches were brought to the Temple before the Sabbath, and after those who had brought them had left, the Boethusians placed stones on them to prevent their being used, since they thought that the Pharisees would not desecrate the Sabbath by moving the stones. But they did not succeed in their object, for the ʿammei ha-aretz joined forces with the Pharisees, and with or without their knowledge removed the willow-branches from under the stones.

The ʿammei ha-aretz here carried out a notable act of support of the Pharisees explicable only on the basis of the close relations which existed between them and which continued unimpaired despite the caution exercised by the Pharisees in refraining from contact with the ʿammei ha-aretz in certain spheres. In this specific instance there may, it is true, have been an additional reason for the action of the ʿammei ha-aretz in that they themselves were interested in the observance of the commandment of the willow-branch, one that is connected with the festival of Tabernacles, that has a popular character, and is of the kind that appeals to the hearts of ordinary people in all generations. The action of the ʿammei ha-aretz is inexplicable had there been a background of enmity and estrangement between them and the Pharisees, for then they would not have performed an indubitable act of support of the Pharisees whereby they made common cause with the Pharisees and helped them directly.

Here mention should once more be made of the abolition of the impurity of the ʿammei ha-aretz during the three pilgrim festivals. Enabling, as it did, the Pharisees and the ḥaverim to come freely into contact

---

[118] TB Sukkah 43b; and cf. Tosefta Sukkah iii, 1.

[119] TB Sukkah 43b.

[120] An analogous term for Sadducees. The view of those who see in the Boethusians a separate sect or identify them with the Essenes is unacceptable (see Y. M. Grintz, *Peraḳim be-Toledot Bayit Sheni*, Jerusalem, 1969, pp. 105–142; and see also Ben-Zion Katz, *Perushim, Tzeduḳim, Ḳanna'im, Notzerim*, Tel Aviv, 1948, pp. 25–34).

with the ʻammei ha-aretz during a pilgrim festival and abolishing, too, the barrier which existed between them during the year, this halakhah is to be seen as the result of the initiative taken by the Pharisaic Sages to prevent a rupture between those scrupulous in their observance of purity and the ʻammei ha-aretz. Besides enlarging the practical possibilities available to the ʻammei ha-aretz to go on pilgrimage to the Temple and allowing them to stay with those who were particular about purity and who undoubtedly constituted a large percentage of the inhabitants of Jerusalem, this halakhah permitted them to use one another's utensils and to trade with one another.

In addition to this, the Pharisees made an effort to intensify among the ʻammei ha-aretz and the nation as a whole the religious-national experience of the pilgrimage. Thus, for example, it was customary to take out the holy vessels and show them to the ʻammei ha-aretz, even if this meant that they touched them. But while the impurity of the ʻammei ha-aretz was indeed abolished during the festival itself, after it, as we have mentioned above, fears were entertained that the vessels were rendered impure retrospectively, and action was taken to restore them to their purity.[121] Against this background we can understand the incident reported in the Tosefta ומעשה שהטבילו את המנורה והיו הצדוקים אומרין, בואו וראו פרושים שמטבילין מאור הלבנה "And it happened that they immersed the menorah [candlestick] to purify it, whereupon the Sadducees said, Come and see Pharisees immersing the light of the moon."[122]

The episode we have quoted concerning the relations between the Pharisees and the ʻammei ha-aretz can also shed light on the relations between the ḥaverim and the ʻammei ha-aretz, for it could just as well have happened to the ḥaverim. In the next section we shall see that even the ḥaverim, who organized themselves in associations and were more extreme than the Pharisees in their strict observance of the laws relating to tithes and purity, nevertheless did not create a social gap between themselves and the ʻammei ha-aretz. From this in itself we can learn something about the relations, too, between the Pharisees and the ʻammei ha-aretz.

---

121 See above, p. 94.

122 Tosefta Ḥagigah iii, 35; and cf. TJ Ḥagigah iii, 79d, where the reading is מטבילין גלגל חמה "They are immersing the sphere of the sun." And see Lieberman, *Tosefta ki-Peshuṭah*, Seder Moʻed, p. 1336.

4. The Relations between the ʻAmmei ha-Aretz and the Ḥaverim

Numerous halakhot testify to the shunning of the ʻammei ha-aretz by ḥaverim and to the taking of precautions by ḥaverim in order not to come into contact with them. From these halakhot it might be inferred that there was an unbridgeable gulf between the ḥaverim and the ʻammei ha-aretz. In point of fact, however, these halakhot express the opposite trend. For the desire to create a social gulf had no need of the scores of halakhot which laid down when and how the ʻam ha-aretz was to be shunned. Instead this could have been conveyed by means of a single, unambiguous halakhah proclaiming a general prohibition against all contacts and close relations with the ʻammei ha-aretz. The many halakhot which set out in detail and minutely examine the problems relating to contacts with the ʻammei ha-aretz and which declare when something is permitted and when prohibited in themselves testify that even as the ḥaverim did not wish to separate themselves entirely from the normative community, as we have shown above, so they also did not want to estrange themselves completely from the ʻammei ha-aretz.

The possibility of establishing family connections between ḥaverim and ʻammei ha-aretz is particularly striking, as is also the fact that the association recognized a heterogeneous family that included ʻammei ha-aretz without this circumstance necessarily leading to the exclusion of the ḥaver from the association.

The halakhah deals with cases of בת עם הארץ שנישאת לחבר "the daughter of an ʻam ha-aretz who married a ḥaver," אשת עם הארץ שנישאת לחבר "the wife of an ʻam ha-aretz who married a ḥaver," עבד עם הארץ שנמכר לחבר "the slave of an ʻam ha-aretz who was sold to a ḥaver."[123] While they indeed צריכין לקבל עליהן וכתחילה "had first to take upon themselves [the obligations of a ḥaver]," there was no prohibition against a ḥaver's marrying a woman who was an ʻam ha-aretz. The halakhah also recognized the possibility of בת חבר שנישאת לעם הארץ "the daughter of a ḥaver who married an ʻam ha-aretz," אשת חבר שנישאת לעם הארץ "the wife of a ḥaver who married an ʻam ha-aretz," and declares that they בחזקתן עד שיחשדו "retain their former status until there is reason to suspect them," or, in the view of R. Simeon b. Eleazar, צריכין לקבל עליהן כתחילה "they have first to take upon themselves [the obligations

[123] Tosefta Demai ii, 16; and see Tosefta ʻAvodah Zarah iii, 9.

of a ḥaver]."[124] In Tosefta 'Avodah Zarah there are differences of opinion about the status of some of these cases. לוקחין מעם הארץ עבדים ושפחות בין גדולים ובין קטנים, ומוכרין לעם הארץ עבדים ושפחות בין גדולים ובין קטנים, ולוקחים מהן בנות קטנות אבל לא גדולות דברי רבי מאיר. וחכמים אומרים, גדולה ומקבלת עליה. ואין נותנין להן בנות, לא גדולות ולא קטנות, דברי רבי מאיר. וחכמים אומרים, נותן לו גדולה ופוסק עמו על מנת שלא תעשה טהרות על גביו. מעשה ברבן גמליאל הזקן, שהשיא את בתו לשמעון בן נתנאל הכהן, ופסק עמו על מנת שלא תעשה טהרות על גביו "One may buy from an 'am ha-aretz male and female slaves, both adults and minors, one may sell to an 'am ha-aretz male and female slaves, both adults and minors, and one may acquire from them minor but not adult servant girls. This is the view of R. Meir. But the Sages said, [One may acquire from them] an adult [servant girl] and she takes upon herself [the obligations of a ḥaver]. One does not give them servant girls, either adults or minors. Such is the view of R. Meir. But the Sages said, One gives him an adult [servant girl] and stipulates with him that it is on condition that she does not prepare food for him in ritual purity. It was told of Rabban Gamaliel the Elder that he married his daughter to Simeon b. Nethanel the priest and stipulated with him that it was on condition that she was not to prepare food for him in ritual purity."[125] It is thus clear that in general the halakhah contented itself with laying down what precisely had to be done that someone who was a ḥaver in an association could continue his membership in it, that is, could continue scrupulously to fulfil his duties as a ḥaver. About the actual establishment of family ties between ḥaverim and 'ammei ha-aretz there were as a rule no reservations.

The halakhah recognized the existence of a ḥaver's family ties with relatives who did not belong to the association. For example, בן חבר שהיה הולך אצל אבי אמו עם הארץ, אין אביו חושש שמא מאכילו טהרות "If a ḥaver's son was in the habit of visiting his maternal grandfather who was an 'am ha-aretz, his father need not be apprehensive that the latter might feed him food prepared in ritual purity."[126] Thus the two sides, evincing discernment, showed consideration for each other. The halakhah goes on to say that אם בידוע שמאכילו טהרות, אסור, ובגדיו טמאין מדרס

124 Tosefta Demai ii, 17; and cf. TB Bekhorot 30b; TB 'Avodah Zarah 30b. On the expression כתחילה in these halakhot, see Epstein, *Mavo le-Nusaḥ ha-Mishnah*, II, pp. 1260 ff.; Lieberman, *Tosefta ki-Peshuṭah*, Seder Zera'im, p. 218.

125 Tosefta 'Avodah Zarah iii, 9–10; and see Lieberman, *Tosefet Ri'shonim*, II, pp. 190–191.

126 Tosefta Demai ii, 15; iii, 5; TB Yevamot 114a.

"if it is known that he feeds him food prepared in purity, it is prohibited [for him to visit his grandfather], and his garments are midras-impure." Here the halakhah refers to a clear instance of the defilement of a ḥaver by a member of his family. Yet even in such an instance the halakhah did not call for the severance of family ties nor did it prescribe any punishment for the ḥaver who continued to maintain such family ties although they necessarily led to his impurity. It is indeed probable that in this case some concession was made since it concerned a ḥaver's son, and that such a concession would not have been made to the ḥaver himself. Yet the absence of any punishment, of the severance of relations with the ḥaver's family, and the lack of any condemnation of the ties with the ʻam ha-aretz, these in themselves testify to the desire on the part of the association to preserve the family connections with non-ḥaverim even if it involved the danger of ṭevel or impurity.

In their choice of places of residence the ḥaverim did not shun any settlement, as did, for example, the Dead Sea sect. Nor did they live segregated in the settlement itself, so that their homes could be found in close proximity to those of the ʻammei ha-aretz. Thus, for example, we read in talmudic literature גגו של חבר למעלה מגגו של עם הארץ "If the ḥaver's roof is higher than that of the ʻam ha-aretz," גגו של חבר בצד גגו של עם הארץ "If the ḥaver's roof adjoined that of the ʻam ha-aretz,"[127] or חבר ועם הארץ שהיו שרויין בחצר "A ḥaver and an ʻam ha-aretz who lived in a courtyard,"[128] and הפנימית של חבר והחיצונה של עם הארץ "[If there are two courtyards one within the other,] the inner belonging to a ḥaver and the outer to an ʻam ha-aretz,"[129] and so on. It is clear that living together in such close proximity could not have been maintained against a background of enmity and estrangement.

Mention should be made of the fact that an ʻam ha-aretz could be accepted as a ḥaver in an association. Although the stages of admission to an association combined to form a lengthy process, it is an extremely significant circumstance that the halakhah refers to an ʻam ha-aretz's passing through this process and to his being accepted as a full ḥaver in an association. Thus the halakhah speaks several times of עם הארץ שקיבל עליו דברי חבירות "an ʻam ha-aretz who took upon himself to ob-

---

127 Tosefta Ṭohorot ix, 11; TB ʻAvodah Zarah 70b.

128 Tosefta Ṭohorot ix, 1.

129 TB ʻAvodah Zarah 70b.

serve the obligations of a ḥaver,''[130] or עם הארץ שנתמנה להיות חבר ''an 'am ha-aretz who was made a ḥaver.''[131]

Very many halakhot deal with the problems relating to the extent to which a ḥaver had to be careful in his contacts with the 'ammei ha-aretz so as not to transgress with ṭevel, demai, or impurity. As previously indicated, these halakhot testify to close relations, even as they attest to an estrangement, between them. In the halakhot of this type we find several expressions which reveal the existence of friendly ties between ḥaverim and 'ammei ha-aretz in matters relating to daily life. Although not all such halakhot mention specifically a ḥaver, it is clear from the context that they deal with ḥaverim and with 'ammei ha-aretz.

The conditions of admission to an association state that it was prohibited for a ḥaver to be the guest of an 'am ha-aretz, nor was he allowed to have an 'am ha-aretz as a guest in his own home.[132] This obligation, which was, it is clear, not always strictly observed in daily life, was even in the context of the conditions of admission to the association the subject of dispute. Thus R. Judah maintained that אף המתארח אצל עם הארץ נאמן ''one who is the guest of an 'am ha-aretz may yet be considered a ne'eman.''[133] That this opinion of R. Judah was not a later halakhah which bore no relation to the situation prevailing in the associations can be seen from the fact that he quoted a tradition relating to associations on this very subject. תני, אמר רבי יודה מימיהן של בעלי בתים לא נמנעו להיות מתארחין אצל בעלי בתים חביריהם, אף על פי כן נוהגין היו בפירותיהן מתוקנין לתוך בתיהן ''It has been taught, R. Judah said, Householders never refrained from being the guests of householders who were their friends [and were 'ammei ha-aretz], and despite this [they continued to be regarded as ne'emanim, since] they tithed their fruits properly in their own homes.''[134] This tradition, stated so decisively, leaves no room for doubt and shows clearly that in practice the ḥaverim did not refrain from being the guests of 'ammei ha-aretz. At the same time, there is no doubt that the ḥaverim did their utmost, while guests of the 'ammei ha-aretz, to guard against transgressing through ṭevel or impurity. This is reflected in various halakhot, such as the following.

130 Tosefta Demai ii, 3, 5; TB Bekhorot 30b.

131 TJ Ḥagigah iii, 79c; TJ 'Eruvin ix, 25d.

132 See above, pp. 125–131, for the conditions of admission, and pp. 131–135, for the obligations of a ḥaver.

133 Demai ii, 2.

134 TJ Demai ii, 22d; and cf. Tosefta Demai ii, 2. And see Lieberman, *Tosefta ki-Peshuṭah*, Seder Zera'im, p. 211.

חבר שהיה ישן בתוך ביתו של עם הארץ, וכליו מקופלין ומונחין תחת ראשו, סנדליו וחביתו לפניו, הרי אלו טהורין מפני שהן בחזקת המשתמר "If a ḥaver slept in the house of an ʻam ha-aretz with his clothes rolled up and placed under his head, and his shoes and his jug in front of him, these are pure since they are under the presumption of having been guarded."[135]

There are also halakhot which deal with the possibility of the ʻammei ha-aretz being guests in the home of a ḥaver and with the problems of impurity likely to result from this. For example, חבר שאמר לעם הארץ צא וישן בתוך הבית, כל הבית ברשותו, צא ישן על מטה פלונית, אינה טמא אלא אותה מטה, ועד מקום שהוא יכול לפשוט את ידו וליגע מאותה מטה. אמר לו שמור לי פרה זו שלא תכנס בתוך הבית, שמור לי פרה זו שלא תשבר את הכלים הרי אלו טהורין שלא מסר לו אלא שמירת פרה בלבד; אבל אם אמר לו שמור לי הבית הזה שלא תכנוס לתוכו פרה, שמור לי כלים הללו שלא תשברם פרה, הרי אלו טמאין "If a ḥaver said to an ʻam ha-aretz, Go and sleep in the house, the whole house is at his disposal. Go and sleep in a certain bed, only that bed is impure as well as up to the place that he is able, by stretching out his hand, to touch from that bed. If he said to him, Guard this cow for me, so that it does not go into the house, guard this cow for me, so that it does not break the utensils, these things are pure, since he entrusted him only with guarding the cow. But if he said to him, Guard this house for me, so that the cow does not go into it, guard these utensils for me, so that the cow does not break them, then these things are impure."[136] Thus we have here not a chance visit by a guest, as in instances mentioned earlier, but a case in which a ḥaver asked the ʻam ha-aretz to go and look after his house or his property. Similarly, המניח עם הארץ בתוך ביתו לשומרו, והלה מוול או כפות, בזמן שרואה את הנכנסין ואת היוצאין אינו טמא, אלא עד מקום שהם יכולים לטמא "If a man leaves an ʻam ha-aretz in his house to guard it, even if the latter is unable to move or is tied, as long as he sees those going in and out, it is not impure except up to the place that they are able to defile."[137] These examples once more show that the ʻammei ha-aretz were not suspected of deliberately defiling the ḥaver, his home, or his utensils. In this spirit the halakhah also declares that המוסר מפתחו לעם הארץ הבית טהור, שלא מסר לו אלא שמירת המפתח "if a man hands his key to an ʻam ha-aretz, the house is pure, since he entrusted him only with safeguarding the key."[138]

135 Tosefta Ṭohorot ix, 2.

136 Tosefta Ṭohorot viii, 3.

137 Tosefta Ṭohorot viii, 7. And see Lieberman, *Tosefet Ri'shonim*, IV, p. 84.

138 Ṭohorot vii, 1; and cf. Tosefta Ṭohorot viii, 1; TB ʻAvodah Zarah 70b.

Whereas the ‘ammei ha-aretz presumably did not partake of the meals of ḥaverim, there are halakhot which point to the possibility of a ḥaver's eating in the home of an ‘am ha-aretz. חבר שהיה מיסב במשתה וסעודת עם הארץ, אפילו רואין אותו נוטל ואוכל מיד, נוטל ושותה מיד, אין זו חזקה למעשרות, שמא עישר בלבו "A ḥaver who partook of the drink or meal of an ‘am ha-aretz, even if he was seen to take and eat immediately, take and drink immediately, this is no presumption that the tithes had been separated, for he may have separated them in thought."[139]

The halakhah also deals with food and utensils which passed from the possession of an ‘am ha-aretz into that of a ḥaver and from the possession of a ḥaver into that of an ‘am ha-aretz, and as usual concerns itself with the resulting problems relating to the separation of tithes, the Sabbatical Year, the observance of purity, and so on. Thus there are halakhot which refer to חבר שלקח עיסה מעם הארץ "a ḥaver who took dough from an ‘am ha-aretz," or חבר שלקח פירות מעם הארץ "a ḥaver who took produce from an ‘am ha-aretz."[140] So, too, reference is made to the depositing of produce with an ‘am ha-aretz. For example, המפקיד פירות אצל עם הארץ, הרי הן כחזקתן למעשרות ולשביעית "If a man deposits produce for safekeeping with an ‘am ha-aretz, it retains its status as regards tithes and the Sabbatical Year."[141] Yet here also the halakhah lays down certain limitations, such as מפקידין תרומה אצל ישראל עם הארץ, ואין מפקידין תרומה אצל כהן עם הארץ, מפני שלבו גס בה "One may deposit terumah for safekeeping with an Israelite who is an ‘am ha-aretz, but one may not deposit terumah for safekeeping with a priest who is an ‘am ha-aretz, since he might take liberties with it."[142]

Between the ḥaverim and the ‘ammei ha-aretz there developed not only neighbourly and friendly, but also business and "professional," relations. So, for example, עם הארץ שהיה משתמש בחנות חבר, אף על פי שחבר יוצא ונכנס מותר, אינו חושש שמא מחליף "If an ‘am ha-aretz uses a ḥaver's shop, although the ḥaver goes in and out, [everything in the shop] is permitted, and there is no fear that he has exchanged things,"[143] or חבר שהיה חוכר מעם הארץ "A ḥaver who hired from an ‘am ha-aretz a field for which the annual rent was to be paid in kind..."[144] Mention has previously been made of the halakhah which permits male and

139 Tosefta Demai iii, 7; etc.

140 TJ Demai i, 22a.

141 Tosefta Demai iv, 22; and cf. TJ Demai ii, 22c.

142 Tosefta Demai iv, 28; TB Giṭṭin 61b.

143 Tosefta Demai iii, 9.

144 Tosefta Demai iii, 5.

female slaves to be bought from and even sold to an 'am ha-aretz. לוקחין מעם הארץ עבדים ושפחות בין גדולים ובין קטנים, ומוכרין לעם הארץ עבדים ושפחות בין גדולים ובין קטנים "One may buy from an 'am ha-aretz male and female slaves, both adults and minors, and one may sell to an 'am ha-aretz male and female slaves, both adults and minors."[145] The halakhah also deals with the case of רופא חבר שהיה מאכיל לחולה עם הארץ "a physician who was a ḥaver and fed a sick person who was an 'am ha-aretz."[146]

In most of the instances cited the ḥaver either took the initiative in coming into contact with the 'am ha-aretz or was in a superior position to him. But there were also cases in which the ḥaver had need of the 'am ha-aretz, as for example, הטוחן אצל עם הארץ ואצל הכותי, אינו חושש משום טומאה "He who grinds [his produce] with an 'am ha-aretz or with a Cuthaean has no reason to be apprehensive of impurity."[147] A notable instance in this connection is the statement בן חבר שלמד אצל עם הארץ, עבד חבר שלמד אצל עם הארץ, הרי הן בחזקתן עד שיחשדו "The son of a ḥaver who studied under an 'am ha-aretz, the slave of a ḥaver who studied under an 'am ha-aretz, retain their former status until there are grounds for suspecting them."[148] The reference here is apparently to a case in which the 'am ha-aretz taught the ḥaver's son or slave a trade, although the possibility of his teaching them the Torah is not to be excluded.

This last case points to the existence of relations between a ḥaver's family and slaves and an 'am ha-aretz. Particularly large is the number of halakhot dealing with the relations which prevailed between the wife of a ḥaver and that of an 'am ha-aretz and which arose from a desire for peace and from neighbourly and friendly ties.

Even as we find the 'am ha-aretz a guest in the ḥaver's home, so are there many instances of the wife of an 'am ha-aretz entering the home of a ḥaver, as for example, אשת חבר שהניחה לאשת עם הארץ טוחנת בתוך ביתה, פסקה הרחים הבית טמא, לא פסקה הרחים אין טמא, אלא עד מקום שהיא יכולה לפשוט את ידה ולגע. היו שתים, בין כך ובין כך הבית טמא, שאחת טוחנת ואחת ממשמשת, דברי רבי מאיר, וחכמים אומרים, אין טמא אלא עד מקום שהן יכולין לפשוט את ידן ולגע "If the wife of a ḥaver left the wife of an 'am

145 Tosefta 'Avodah Zarah iii, 9.

146 TJ Demai iii, 23b.

147 Tosefta Demai iv, 27.

148 Tosefta Demai ii, 18. There we also find the reverse possibility of בן עם הארץ הלמד אצל חבר וכו' "the son of an 'am ha-aretz who studied under a ḥaver..."

ha-aretz grinding corn in her house, the house is deemed impure if she ceased to turn the handmill, but if she did not cease to turn the handmill only that part of the house is deemed impure which, by stretching out her hand, she can touch. If there were two women, the house is in any case impure, since while the one is grinding, the other can go about touching. This is the view of R. Meir. But the Sages said, Only that part of the house is impure which, by stretching out their hands, they can touch."[149] In this instance, too, despite the possibility that she might defile the ḥaver's house and utensils, the 'am ha-aretz's wife was not prohibited in advance from going into the ḥaver's house. Instead the circumstances under which she causes impurity are set forth. In this case, as in many others, we once more find R. Meir adopting the stricter view in the halakhot relating to ḥaverim and to purity. Another halakhah on the same subject states אשתו של עם הארץ טוחנת עם אשת חבר בזמן שהיא טמאה, אבל בזמן שהיא טהורה לא תטחן, ר׳ שמעון אומר, בזמן שהיא טמאה לא תטחון, שאף על פי שאינה אוכלת נותנת לאחרות ואוכלות "The wife of an 'am ha-aretz may grind along with the wife of a ḥaver when she is impure, but not when she is pure. R. Simeon said, Even when she is impure she should not grind with her, for although she herself does not eat, she gives to other women and they eat."[150] The view of the Tanna mentioned first in this halakhah is apparently based on the assumption that the wife of an 'am ha-aretz can be relied upon, when impure with menstrual-impurity, to be careful not to defile anything in her immediate surroundings, and hence the ḥaver's wife may grind with her. But when not menstruous, the wife of an 'am ha-aretz is not careful, and the ḥaver's wife should accordingly not grind along with her.

There is an halakhah which permits the wife of a ḥaver to lend utensils to the wife of an 'am ha-aretz in certain cases. אשת חבר משאלת לאשת עם הארץ, נפה וכברה, ובוררת וטוחנת ומרקדת עמה, אבל משתטיל את המים לא תגע אצלה, שאין מחזיקין ידי עוברי עברה. וכלן לא אמרו אלא מפני דרכי שלום "The wife of a ḥaver may lend the wife of an 'am ha-aretz a winnow and a sieve, and may even winnow and grind corn or sift flour with her. But once she pours water [over the flour] she should not touch [what is] with her [the 'am ha-aretz's wife], for no help must be given to those committing a transgression. All these things were only per-

---

149 Ṭohorot vii, 4; and cf. viii, 5.

150 Tosefta Ṭohorot viii, 4; and cf. TB Giṭṭin 61b–62a; and see Lieberman, *Tosefet Ri'shonim*, IV, pp. 81–82.

mitted in the interests of peace."[151] Here, too, we have one of the reasons for maintaining good relations between ḥaverim and the ʿammei ha-aretz in so far as this was possible, namely, "in the interests of peace."

[151] Shevi'it v, 9; Giṭṭin v, 9.

CHAPTER FIVE

THE RELATIONS BETWEEN THE 'AMMEI HA-ARETZ AND THE TALMIDEI ḤAKHAMIM

1. Introductory Remarks

The 'am ha-aretz la-Torah, who is contrasted with the talmid ḥakham, is referred to in statements and halakhot dating from after the destruction of the Second Temple onwards.

In the period of Jabneh and of Usha, as also in the days of R. Judah ha-Nasi, there occur statements and halakhot which contain harsh expressions condemning the 'am ha-aretz le-mitzvot. As we have seen, the ḥaverim did not attempt to create a barrier between themselves and the 'ammei ha-aretz, whom they shunned only in so far as this was necessary in order not to transgress through ṭevel,[a] demai, and impurity. As against this, the talmidei ḥakhamim, so one gains the impression, insisted on widening the social gap between themselves and the 'ammei ha-aretz.

In their social and practical application there was apparently no great difference between the concepts of the 'am ha-aretz le-mitzvot and the 'am ha-aretz la-Torah. The social stratum, which neglected the commandments relating to the tithes and to purity and which was contrasted in the days of the Second Temple with the ḥaverim in the associations, was presumably more or less identical in its social composition with the social stratum which, contrasted with the talmidei ḥakhamim, did not join in the process that came into being after the destruction of the Second Temple with the aim of setting the Torah and its study at the centre of Jewish life.

Contrasted with the 'am ha-aretz, the term ḥaver, denoting esteem and indicating a high status, was also used after the destruction of the Second Temple to refer to a talmid ḥakham. So, for example, **ואין חברים אלא תלמידי חכמים** "ḥaverim are none other than talmidei ḥakhamim."[1] Accordingly in the generations following the destruction of the Second

---

[1] Bava Batra 75a.

a) Produce at the stage at which tithes and terumot should be, but have not yet been, separated.

Temple the expression ḥaver occurs in numerous halakhot and statements in the sense of a talmid ḥakham.[2] The sources also have the expression חבירי תורה "ḥaverim of the Torah,"[3] while the Eretz Israel Amoraim were referred to in Babylonia by the term חבורה "the association."[4] [5]

From the association of Second Temple times the talmidei ḥakhamim inherited not only the title of ḥaver but also something of the framework of the association. The study of the Torah was frequently carried on in an association consisting of teacher and pupils, and at times such an association would move from place to place for the study of the Torah. It is against this background that we have to view the statement עשו כתות כתות ועסקו בתורה, לפי שאין התורה נקנית אלא בחבורה "Form groups and occupy yourselves in the Torah, for the knowledge of the Torah is acquired only in an association."[6] In such a framework there was clearly a communal life, in which the pupils ministered to the Sages, and hence the view becomes intelligible which declares that אפילו קרא ושנה ולא שמש תלמידי חכמים הרי זה עם הארץ "even one who has learnt Scripture and Mishnah but has not ministered to the Sages is an 'am ha-aretz."[7]

These associations, and especially those in which the Sages moved from place to place with their pupils, were presumably accustomed to have meals in common and perhaps even pooled their money. In talmudic literature the relations between teacher and pupil are likened to those between brothers or between father and son.[8] Against this background we can understand why the injunction that אל יסב בחבורה של

---

[2] For example, TJ Mo'ed Ḳaṭan iii, 81d; TB Bekhorot 36a; TB 'Eruvin 32b; Avot de-R. Nathan, Version B, xxxi, ed. Schechter, p. 68; Midrash Proverbs vi, ed. Buber, p. 56.

[3] TJ Berakhot i, 2d; Tanḥuma, Nitzavim, iv.

[4] TB Shabbat 3a.

[5] W. Bacher, "Zur Geschichte der Schulen Palästina's im 3. und 4. Jahrhundert: Die Genossen (חברייא)," *MGWJ*, 43 (1899), pp. 345–360, maintains that a scholar, who had not yet received the title of Rabbi, was called a ḥaver. This view is however unacceptable.

[6] TB Berakhot 63b. And see A. Büchler, "Learning and Teaching in the Open Air in Palestine," *JQR* (N.S.), 4 (1913–1914), pp. 485–491.

[7] TB Berakhot 47b.

[8] For example, TB 'Eruvin 73a. And see G. Allon, *Toledot ha-Yehudim be-Eretz Yisra'el bi-Teḳufat ha-Mishnah ve-ha-Talmud*, Tel Aviv, I: 1958³, II: 1961² — see I, pp. 294–323, who compares the communal life which sometimes prevailed among the Sages with the usages adopted by Jesus and his disciples, among whom there were also "pupils ministering to a Sage," meals in common, and so on.

עמי הארץ "he shall not take a set meal in the company of ʿammei ha-aretz"[9] should have been included in the list of things unbecoming to a talmid ḥakham.

2. Manifestations of Hatred between Talmidei Ḥakhamim and ʿAmmei ha-Aretz

Contrary to the ḥaverim who generally were not opposed to "mixed marriages" between themselves and the ʿammei ha-aretz, the Sages expressed resolute opposition to such unions. תנו רבנן לעולם ימכור אדם כל מה שיש לו וישא בת תלמיד חכם, שאם מת או גולה, מובטח לו שבניו תלמידי חכמים, ואל ישא בת עם הארץ, שאם מת או גולה בניו עמי הארץ. תנו רבנן לעולם ימכור אדם כל מה שיש לו וישא בת תלמיד חכם, וישיא בתו לתלמיד חכם, משל לענבי הגפן בענבי הגפן דבר נאה ומתקבל, ולא ישא בת עם הארץ, משל לענבי הגפן בענבי הסנה דבר כעור ואינו מתקבל. תנו רבנן לעולם ימכור אדם כל מה שיש לו וישא בת תלמיד חכם, לא מצא בת תלמיד חכם, ישא בת גדולי הדור, לא מצא בת גדולי הדור, ישא בת ראשי כנסיות, לא מצא בת ראשי כנסיות ישא בת גבאי צדקה, לא מצא בת גבאי צדקה ישא בת מלמדי תינוקות — ולא ישא בת עמי הארץ, מפני שהן שקץ ונשותיהן שרץ, ועל בנותיהן הוא אומר "ארור שוכב עם כל בהמה"... תניא היה רבי מאיר אומר, כל המשיא בתו לעם הארץ כאילו כופתה ומניחה לפני ארי, מה ארי דורס ואוכל ואין לו בושת פנים, אף עם הארץ מכה ובועל ואין לו בושת פנים "Our Rabbis taught, A man should always sell all he has and marry the daughter of a talmid ḥakham, for if he dies or goes into exile he is assured that his children will be talmidei ḥakhamim, and he should not marry the daughter of an ʿam ha-aretz, for if he dies or goes into exile his children will be ʿammei ha-aretz. Our Rabbis taught, A man should always sell all he has and marry the daughter of a talmid ḥakham and marry his daughter to a talmid ḥakham. This may be compared to bunches of grapes combined with bunches of grapes, which is a seemly and acceptable thing. But he should not marry the daughter of an ʿam ha-aretz. This may be compared to bunches of grapes combined with the berries of a thorn bush, which is a repulsive and unacceptable thing.

Our Rabbis taught, A man should always sell all he has and marry the daughter of a talmid ḥakham. If he cannot find a daughter of a talmid ḥakham, he should marry the daughter of [one of] the great men of the generation. If he cannot find a daughter of [one of] the great men of the generation, he should marry the daughter of the head

[9] TB Berakhot 43b; and cf. Massekhet Derekh Eretz iii, 1, ed. Higger, p. 97.

of synagogues. If he cannot find the daughter of the head of synagogues, he should marry the daughter of a charity treasurer. If he cannot find a daughter of a charity treasurer, he should marry the daughter of an elementary school teacher, but he should not marry the daughter of ʻammei ha-aretz, for they are detestable and their wives are vermin, and of their daughters it is said, ʻCursed be he who lies with any kind of beast.' ...It was taught, R. Meir used to say, Whoever marries his daughter to an ʻam ha-aretz is as though he bound and laid her before a lion, for just as a lion tears [his prey] and devours it and has no shame, so an ʻam ha-aretz strikes and cohabits and has no shame."[10]

These Baraitot include, on the one hand, logical arguments against "mixed" families, such as, for example, that if a man marries the daughter of a talmid ḥakham he can be assured that, even should he die or go into exile, his children will be educated in the Torah; on the other hand, they contain harsh curses and irrational statements, such as, that if a man marries the daughter of ʻammei ha-aretz, he associates with "detestable things and with vermin."

Whereas we have found ḥaverim and ʻammei ha-aretz living in close proximity to one another, the Sages warned against dwelling near an ʻam ha-aretz. אמר רבי שמעון בן לקיש, אם תלמיד חכם נוקם ונוטר כנחש הוא, חגריהו על מתניך, אם עם הארץ הוא חסיד, אל תדור בשכונתו "R. Simeon b. Laḳish said, [Even] if a talmid ḥakham is vengeful and bears malice like a serpent, gird him to your loins, [whereas even] if an ʻam ha-aretz is pious, do not dwell in his vicinity."[11] This statement is to be connected with the general demand, which we shall later discuss, to shun the ʻammei ha-aretz. Against this background we can understand the remark of R. Dosa b. Harkinas that ...וישיבת בתי כנסיות של עמי הארץ, מוציאין את האדם מן העולם "...sitting in the synagogues of the ʻammei ha-aretz puts a man out of the world."[12] From the period of Jabneh, prayer and the synagogue came to occupy a progressively central place in the religious life of the community. Presumably the ʻammei ha-aretz did not fully comprehend the purpose of the Sages, so that their synagogues were not places of genuine prayer and of the study of the Torah. It is on the basis of this that the following statement is to be understood. תניא ר׳ ישמעאל בן אלעזר אומר בעון שני דברים עמי הארצות מתים, על

[10] TB Pesaḥim 49a–b; and cf. Pirḳa de-Rabbenu ha-Ḳadosh, Bava de-Arba'ah, ed. Schönblum, 21b.

[11] TB Shabbat 63a.

[12] Avot iii, 10. For the alternative reading כנסיות "assemblies," see above, p. 21, note 76.

שקורין לארון הקודש ארנא, ועל שקורין לבית הכנסת בית עם "It was taught, R. Ishmael b. Eleazar said, On account of two things the 'ammei ha-aretz die, because they call the holy ark a chest, and because they call the synagogue a bet 'am [a people's house]."[13]

Several halakhot and statements prohibit a talmid ḥakham from associating and becoming friendly with 'ammei ha-aretz because of their ignorance of the Torah and their general untrustworthiness. For example, תנו רבנן ששה דברים גנאי לו לתלמיד חכם, אל יצא כשהוא מבושם לשוק, ואל יצא יחידי בלילה, ואל יצא במנעלים המטולאים, ואל יספר עם אשה בשוק, ואל יסב בחבורה של עמי הארץ, ואל יכנס באחרונה לבית המדרש וכו׳ "Our Rabbis taught, Six things are unbecoming to a talmid ḥakham. He should not go abroad scented, he should not go out at night alone, he should not go abroad in patched sandals, he should not converse with a woman in the street, he should not take a set meal in the company of 'ammei ha-aretz, and he should not be the last to enter the bet ha-midrash [the house of study]..."[14] Another statement sets forth the "diminishing value" of a talmid ḥakham, the friendlier he becomes with an 'am ha-aretz. — חבר בפני עם הארץ, ככלי זהב. השיח עמו, ככלי זכוכית. אכל ושתה עמו ככלי חרס "A ḥaver [talmid ḥakham] is regarded by an 'am ha-aretz as a golden vessel; if he converses with him, as a glass vessel; if he eats and drinks with him, as an earthen vessel."[15]

The low opinion that the Sages had of the 'am ha-aretz is expressed in the following Baraita. תנו רבנן, ששה דברים נאמרו בעמי הארץ, אין מוסרין להן עדות, ואין מקבלין ממנו עדות, ואין מגלין להן סוד, ואין ממנין אותן אפוטרופוס על היתומים, ואין ממנין אותן אפוטרופוס על קופה של צדקה, ואין מתלוין עמהן בדרך, ויש אומרים אף אין מכריזין על אבידתו "Our Rabbis taught, Six things were said of the 'ammei ha-aretz. We do not commit testimony to them, we do not accept testimony from them, we do not reveal a secret to them, we do not appoint them guardians of orphans, we do not appoint them stewards of charity funds, and we do not join their company on the road. Some say, We also do not proclaim their losses."[16] The

---

13 TB Shabbat 32a. In MSS. and in the earlier authorities the author of the statement is given as R. Simeon b. Eleazar, which seems more probable. And see Dikdukei Soferim.

14 TB Berakhot 43b; and cf. Derekh Eretz v, 1, ed. Higger, p. 116.

15 Avot de-R. Nathan, Version B, xxxi, ed. Schechter, p. 68. The Halberstam MS. has השיח עמו שיחה בטלה "if he engages in idle conversation with him."

16 TB Pesaḥim 49b; and cf. אמר רבי אלעזר, עם הארץ אסור להתלוות עמו בדרך, שנאמר "כי היא חייך ואורך ימיך," על חייו לא חס על חיי חבירו לא כל שכן "R. Eleazar said, One must not join company with an 'am ha-aretz on the road, since it is said,

practical significance of this halakhah is that it places the ‘am ha-aretz on a level with irresponsible, untrustworthy people who, like a deaf mute, an insane person, and a minor, are disqualified from giving evidence. The ‘ammei ha-aretz were also compared to non-Jews. רבי יהודה אומר, עם הארץ הרי הוא כגוי "R. Judah said, An ‘am ha-aretz is like a gentile."[17]

The Sages did not consider it possible for an ‘am ha-aretz to perform good deeds or have outstanding qualities. Thus there is the statement in the name of R. Simeon b. Yoḥai that עם הארץ, אפילו חסיד, אפילו ישרן, אפילו קדוש ונאמן – ארור הוא לאלוהי ישראל "an ‘am ha-aretz, even if he is pious, even if he is upright, even if he is holy and trustworthy, cursed be he unto the God of Israel."[18] The Sages did not even permit the Torah to be studied in the presence of ‘ammei ha-aretz. תנא רבי חייא, כל העוסק בתורה לפני עם הארץ, כאילו בועל ארוסתו בפניו, שנאמר "תורה צוה לנו משה מורשה," אל תקרי מורשה אלא מאורסה "R. Ḥiyya taught, Whoever studies the Torah in front of an ‘am ha-aretz is as though he cohabited with his betrothed in his presence, for it is said, ‘Moses commanded us a Torah, as a possession [morashah] for the assembly of Jacob.’ Do not read ‘morashah’ but ‘me’orasah,’ betrothed."[19] R. Judah ha-Nasi is the author of the statement that אין פורענות בא לעולם אלא בשביל עמי הארץ "it is the ‘ammei ha-aretz who bring misfortune on the world."[20] He also expressed opposition to the ‘am ha-aretz's eating meat. תניא רבי אומר, עם הארץ אסור לאכול בשר (בהמה), שנאמר "זאת תורת הבהמה והעוף," כל העוסק בתורה מותר לאכול בשר בהמה ועוף, וכל שאינו עוסק בתורה אסור לאכול בשר בהמה ועוף "It was taught, Rabbi [R. Judah ha-Nasi] said, An ‘am ha-aretz may not eat the flesh of cattle, for it is said, ‘This is the law [Torah] pertaining to beast and bird.’ Whoever engages in the study of the Torah may eat the flesh of beast and bird, but whoever does

---

‘For that [the Torah] means life to you and length of days.’ [Seeing that] he has no consideration for his own life, how much more [has he no consideration] for the life of his companion" (*ibid.*). Cf. Kallah Rabbati ii, 14, ed. Higger, p. 210. In Seder Eliyahu Rabbah xiii, ed. Friedmann, pp. 59–61, it is emphasized at length that one should not eat often with ‘ammei ha-aretz.

17 Pirḳa de-Rabbenu ha-Ḳadosh, Bava de-Arba‘ah, ed. Schönblum, 21b. Although Pirḳa de-Rabbenu ha-Ḳadosh was apparently redacted at a late date, the statements relating to the ‘am ha-aretz correspond for the most part to the Baraitot in TB Pesaḥim 49a–b, and are therefore presumably authentic.

18 Pirḳa de-Rabbenu ha-Ḳadosh, *loc. cit.*

19 TB Pesaḥim 49b.

20 TB Bava Batra 8a.

not engage in the study of the Torah may not eat the flesh of beast and bird."[21]

The extreme of harshness was reached in the following statements. אמר רבי אלעזר, עם הארץ מותר לנוחרו ביום הכיפורים שחל להיות בשבת, אמרו לו תלמידיו, רבי אמור לשוחטו, אמר להן, זה טעון ברכה וזה אינו טעון ברכה "R. Eleazar said, An 'am ha-aretz may be stabbed even on a Day of Atonement which falls on a Sabbath. Said his pupils to him, Master, say rather, may be ritually slaughtered. He replied, The latter requires a benediction, the former does not."[22] Similarly, אמר רבי שמואל בר נחמני אמר רבי יוחנן, עם הארץ מותר לקורעו כדג, אמר רבי שמואל בר יצחק, ומגבו "R. Samuel b. Naḥmani said in the name of R. Johanan, One may tear an 'am ha-aretz like a fish. Said R. Samuel b. Isaac, And [this means] along his back."[23]

By the nature of things we read in talmudic literature more about the attitude of talmidei ḥakhamim to the 'ammei ha-aretz than about the attitude of the 'ammei ha-aretz to talmidei ḥakhamim. Since no literature of the 'ammei ha-aretz is extant, it is difficult to give a comprehensive picture of their expressions and thoughts about, and their attitude to, talmidei ḥakhamim.

There were Sages who testified to the attitude of the 'ammei ha-aretz to them. תניא רבי אליעזר אומר, אילמלא אנו צריכין להם למשא ומתן, היו הורגין אותנו "It was taught, R. Eliezer said, Were it not that we are necessary to them for trade, they would kill us." There is similarly the statement of R. Ḥiyya that גדולה שנאה ששונאין עמי הארץ לתלמיד חכם, יותר משנאה ששונאין עכו"ם את ישראל ונשותיהן יותר מהן "greater is the hatred wherewith the 'ammei ha-aretz hate a talmid ḥakham than the hatred wherewith the heathens hate Israel, and their wives [hate even] more than they."[24] It is against this background that we must view R. Jose's remark והשונא חכמים ותלמידיהם... אין לו חלק לעולם הבא "And he who hates Sages and their pupils... has no share in the world to come."[25]

Confirmation of these statements is to be found in the well-known declaration of R. Akiva about the period when he was still an 'am ha-aretz. תניא אמר רבי עקיבא, כשהייתי עם הארץ, אמרתי מי יתן לי תלמיד חכם

---

21 TB Pesaḥim 49b.

22 *Ibid.* The Munich MS. quotes it in the name of R. Eliezer. And see Diḳduḳei Soferim.

23 TB Pesaḥim 49b.

24 *Ibid.*

25 Derekh Eretz Rabbah xi, ed. Higger, pp. 313–314.

ואנשכנו כחמור, אמרו לו תלמידיו, רבי אמור ככלב, אמר להן, זה נושך ושובר עצם, וזה נושך ואינו שובר עצם "It was taught, R. Akiva said, When I was an 'am ha-aretz I said, Would that I had a talmid ḥakham before me, and I would bite him like an ass. Said his pupils to him, Master, say rather like a dog. He answered them, The former bites and breaks a bone, the latter bites but does not break a bone."[26]

Several incidents reflect the atmosphere which characterized the above statements. רבי מאיר הוה יליף דריש בכנישתא דחמתא כל לילי שובא והוה תמה חדא איתתא יליפה שמעה קליה, חד זמן עני דריש אזלת בעית מיעול לביתיה ואשכחת בוצינא מיטפי, אמר לה בעלה, הן הוייתה, אמרה ליה, מישמעא קליה דדרושא, אמר לה, מכך וכך דלית ההיא איתתא עללה להכא לבייתה עד זמן דהיא אזלה ורקקה גו אפוי דדרושא. צפה רבי מאיר ברוח הקודש ועבד גרמיה חשש בעייניה, אמר כל איתתא דידעה מילחוש לעיינה תיתי תילחוש. אמרין לה מגירתא הא עניתיך תיעלין לביתיך עבדי גרמיך לחשה ליה ואת רקקה גו עייניה. אתת לגביה, אמר לה, חכמה את מילחוש לעיינא. מאימתיה עליה, אמרה ליה לא, אמר לה ורוקקין בגויה שבע זימנין והוא טב ליה, מן דרקקת אמר לה, אזלין אמרין לבעליך, חד זמן אמרת לי, והיא רקקה שבעה זימנין. אמרו לו תלמידיו, רבי, כך מבזין את התורה. אילו אמרת לו לא הווית מייתי ליה ומלקין לה ספסליה ומרצייין ומרצייה ליה לאיתתיה. אמר לון, ולא יהא כבוד מאיר ככבוד קונו, מה אם שם הקודש שנכתב בקדושה, אמר הכתוב שיימחה על המים, בשביל להטיל שלום בין איש לאשתו, וכבוד מאיר לא כל שכן.

"R. Meir used to preach in the synagogue of Ḥammeta every Friday evening, and a certain woman there made it a habit to listen to him. On one occasion he preached late, and when she wanted to enter her home she found the light out. Her husband asked her, Where have you been? She answered, I was listening to a preacher, whereupon he said, If that is so, you will never enter this house until you have gone and spat in the preacher's face. Having by means of the Holy Spirit learnt what had happened, R. Meir, pretending that he was suffering from a pain in his eye, declared, Let any woman who is able to cure a sore eye by a charm come and charm it for me. Her neighbours said to her, Here is your chance of getting back to your home again. Make out that you are charming him, and spit in his eye. When she came to him, he said to her, Do you know how to cure an eye by a charm? From

---

26 TB Pesaḥim 49b. There is no foundation for the view of A. Büchler, *Der galiläische 'Am-ha'Areṣ des zweiten Jahrhunderts*, Vienna, 1906, p. 184, that the name Akiva is to be emended to Ya'aḳov or 'Aḳavya so as to transfer the statement from the generation of Jabneh. On the anachronism in the words "Said his pupils to him, Master, say...," see E. E. Urbach, *Ḥazal — Pirḳei Emunot ve-De'ot*, Jerusalem, 1969, p. 572, note 28. For a biography of R. Akiva, see S. Safrai, *R. 'Aḳiva ben Yosef — Ḥayyav u-Mishnato*, Dorot Library, Jerusalem, 1970, pp. 9–33.

fear of him, she said, No. Thereupon he said to her, If you spit in it seven times, it will get better. After she had done so, he said to her, Go, and tell your husband, You told me to spit once, but I have spat seven times. His pupils said to him, Master, should the Torah be made contemptible in this way? Had you told us, would we not have brought in benches and straps and beaten him until he was reconciled with his wife? He said to them, And should not the honour of Meir be like that of his Creator? If the Name of the Lord, written in sanctity, is, so Scripture says, to be blotted out with water, in order that peace may be made between a man and his wife, how much more should Meir's honour be disregarded."[27] A similar incident is recorded in tractate Nedarim. ההוא דאמר לה לדביתהו, קונם שאי את נהנית לי עד שתרוקי בו ברבן שמעון בן גמליאל, אתת ורקק אלבושיה וכו׳ "A man once said to his wife, Konam[a)] that you derive no benefit from me until you spit on Rabban Simeon b. Gamaliel. She went and spat on his garment..."[28] Although the expression 'am ha-aretz is not mentioned in these incidents, they are undoubtedly a living testimony of the hatred of the Sages prevalent among the 'ammei ha-aretz.

The acrimonious character of the statements, which testifies to the seething hatred that existed between the talmidei ḥakhamim and the 'ammei ha-aretz, and the Sages' exhortation to shun and not to rely on them, which was in the nature of excommunicating an entire stratum in the nation — all this has caused surprise and led to difficulties from Gaonic to contemporary times.

Thus the Geonim tried to explain away the harshness of the statements and of the halakhot directed against the 'ammei ha-aretz. One method adopted by them was not to give these remarks a general application but rather to apply them to 'ammei ha-aretz of a certain type or to a particular occasion. So, for example, Sherira Gaon explained

---

27 TJ Soṭah i, 16d. Despite his harsh statements about the 'ammei ha-aretz, R. Meir did not insist on his honour in this case. Moreover when courtesy demanded that respect be shown to the 'ammei ha-aretz, he did so. רבי מאיר חמי אפילו סב עם הארץ ומקים ליה מן קומוי ואמר לא מגן מאריך י״ם "R. Meir, when he saw even an old 'am ha-aretz, would place him before himself, saying, It is not for nothing that he has been granted long life" (TJ Bikkurim iii, 65c; and cf. בישיבה הלך אחר חכמה, במסיבה הלך אחר זקנה "At a session [of the court or of the bet ha-midrash] priority is to be given to wisdom; at a festive gathering, age takes precedence" [TB Bava Batra 120a]).

28 TB Nedarim 66b. See also the episode in which R. Simeon was scrupulous about the honour of Sages (*ibid.*).

a) Ḳonam [a substitute for ḳorban = sacrifice], used in taking a vow of abstinence.

the statement permitting the stabbing of an ʿam ha-aretz on a Day of Atonement which falls on a Sabbath as referring to the case שעם הארץ רודף אחר חבירו להרגו ביום הכיפורים שחל להיות בשבת "of an ʿam ha-aretz who on a Day of Atonement that fell on a Sabbath pursued his neighbour in order to kill him."[29] Such explanations however distort and circumscribe the meaning of these statements. It is of course possible that not every case dealing with the ʿam ha-aretz la-Torah refers to the entire stratum of those who had not studied the Torah, and that the reference is sometimes to those who were drawn to heresy. But even this is not a complete explanation.[30]

Maintaining that the Sages here repeat earlier traditions which echo the attitude of the ancient association to non-ḥaverim, Lieberman[31] and others do not set these statements within the context of the historical situation which obtained at the time they were made. But this view is improbable for the following reasons. i) Statements of this type were made by Sages only after the destruction of the Second Temple and, as far as can be ascertained, were directed against the ʿam ha-aretz la-Torah. ii) We have earlier shown that it was specifically the association of ḥaverim which did not adopt a negative attitude to the ʿammei ha-aretz, did not express itself harshly about them, and did not demand that they be completely shunned. iii) Where sects, such as the Yaḥad, repudiated those who did not belong to them, the shafts of their criticism were directed against the normative community from which they had separated themselves, and not against the ʿammei ha-aretz. Nor presumably would the Sages have taken it upon themselves to be the mouthpiece of traditions of this sort, had such sects still existed.

A solution of the harshness of the statements, in so far as one can be arrived at, is to be found in a detailed examination of their historical and literary background.

As regards the literary background, it should be pointed out that the statements of the Sages are not always to be taken literally. Nor have every Baraita and every statement the obligatory character of a decided halakhah. Some utterances of the Sages were no more than hackneyed expressions. Often derogatory and provocative remarks were made by one talmid ḥakham about another. Of his pupil Levi, R. Judah

---

29 See B. M. Lewin, *Otzar ha-Ge'onim*, Haifa–Jerusalem, 1928–1943, on Pesaḥim 49b, p. 68, and parallel passages, for further explanations in this spirit.

30 See Urbach, *Ḥazal*, p. 573.

31 See S. Lieberman, "The Discipline in the So-Called Dead Sea Manual of Discipline," *JBL*, 71 (1952), pp. 199–206.

ha-Nasi said כמדומה לי שאין לו מוח בקדקדו "It seems to me he has no brains in his head";[32] the Babylonian Amoraim were called בבלאי טפשאי "the stupid Babylonians";[33] and so on.[34] Just as a decided and obligatory halakhah is not to be seen in the statement כל תלמיד חכם שנמצא רבב על בגדו חייב מיתה "Any talmid ḥakham on whose garment a [grease] stain is found is worthy of death,"[35] so statements declaring that it is permitted to kill an 'am ha-aretz are not to be regarded as halakhic. While there is no doubt about the hatred which prevailed at that time between talmidei ḥakhamim and 'ammei ha-aretz, the acrimonious nature of the statements distort and magnify that hatred. It was customary for the Sages to use sharp language, and some of their expressions were current idioms and consequently much more trite and worn than they appear to us today. In the context of the contemporary historical conditions the remarks were also characterized, as we shall see later, by a certain tendentiousness.

Historically, the statements first occurred in the generation of Jabneh, were intensified in the period of Usha, and continued under R. Judah ha-Nasi.

As regards the first occurrence of the statements in the generation of Jabneh following the destruction of the Second Temple, another surprising fact is to be added to the existing ones.

It is specifically the generation of Jabneh that witnessed the spread of the tendency towards equality among the various strata of society, including an equality between talmidei ḥakhamim and those who did not engage in the study of the Torah. In the following statement this idea of social equality finds vigorous expression. מרגלא בפומייהו דרבנן דיבנה, אני בריה וחברי בריה, אני מלאכתי בעיר והוא מלאכתו בשדה, אני משכים למלאכתי והוא משכים למלאכתו, כשם שהוא אינו מתגדר במלאכתי כך אני איני מתגדר במלאכתו, ושמא תאמר אני מרבה והוא ממעיט, שנינו — אחד המרבה ואחד הממעיט, ובלבד שיכוין לבו לשמים "A favourite saying of the Rabbis of Jabneh was, I am God's creature and my fellow is God's creature. My work is in the town and his work is in the country. I rise early for my work and he rises early for his work. Just as he does not presume to do my work, so I do not presume to do his work. Will you say I do

[32] TB Yevamot 9a.

[33] TB Ketubbot 75a; etc.; etc.

[34] See Urbach, *Ḥazal*, pp. 557–569.

[35] TB Shabbat 114a. Cf. also הוא [רבי עקיבא] היה אומר, דלא שימש חכימיא קטלא חייב "He [R. Akiva] used to say, Whoever has not ministered to the Sages is worthy of death" (TJ Nazir vii, 56b).

much and he does little? We have learnt, One may do much or one may do little, it is all the same provided one directs one's heart to Heaven."[36]

This social outlook is undoubtedly the outcome of the general trend in the generation of Jabneh towards reducing dissension, abolishing sects, and demolishing barriers between the various sections of the nation, all with the purpose of sustaining the national existence of the people, now deprived of the Temple.

The idea of equality found expression not only in statements but also in the halakhah, as for example, התוקע לחברו נותן לו סלע, רבי יהודה אומר משום רבי יוסי הגלילי, מנה. סטרו נותן לו מאתים זוז, לאחר ידו נותן לו ארבע מאות זוז. צרם באזנו, תלש בשערו, רקק והגיע בו רוקו, העביר טליתו ממנו, פרע ראש האשה בשוק — נותן ארבע מאות זוז. זה הכלל: הכל לפי כבודו. אמר רבי עקיבא, אפילו עניים שבישראל רואין אותם כאלו הם בני חרין שירדו מנכסיהם, שהם בני אברהם יצחק ויעקב. ומעשה באחד שפרע ראש האשה בשוק, באת לפני רבי עקיבא, וחיבו לתן לה ארבע מאות זוז. אמר לו, רבי תן לי זמן, ונתן לו זמן. שמרה עומדת על פתח חצרה, ושבר את הכד בפניה, ובו כאסר שמן. גלתה את ראשה, והיתה מטפחת ומנחת ידה על ראשה. העמיד עליה עדים ובא לפני רבי עקיבא, אמר לו, רבי, לזו אני נותן ארבע מאות זוז. אמר לו, לא אמרת כלום, החובל בעצמו, אף על פי שאינו רשאי פטור, אחרים שחבלו בו חיבין, והקוצץ את נטיעותיו, אף על פי שאינו רשאי פטור, אחרים שקצצו את נטיעותיו חיבים.

"If a man boxes another man's ear he has to pay him a sela'. R. Judah in the name of R. Jose the Galilaean said, [He has to pay him] a maneh. If he slapped him [on the face] he has to pay him two hundred zuz; if [he did it] with the back of his hand he has to pay him four hundred zuz. If he pulled his ear, plucked his hair, spat so that the spittle reached him, removed his garment from upon him, uncovered the head of a woman in the market place, he must pay four hundred zuz. This is the general principle: it all depends upon the dignity [of the person insulted]. R. Akiva said, Even the poor in Israel are to be considered the children of aristocrats reduced in circumstances, seeing they are the descendants of Abraham, Isaac, and Jacob. It once happened that a certain man uncovered the head of a woman in the market place and when she came before R. Akiva, he ordered the offender to pay her four hundred zuz. The latter said to him, Rabbi, allow me time [in which to carry out the judgment]. R. Akiva agreed and fixed a time for him. Watching her until he saw her standing outside the door of her courtyard, he broke in her presence a pitcher in which there was

---

[36] TB Berakhot 17a.

oil of the value of an isar, whereupon she uncovered her head, collected the oil with her palms, and putting her hands upon her head [anointed it]. He set up witnesses against her and came to R. Akiva and said to him, Rabbi, have I to give such a woman four hundred zuz? R. Akiva said to him, Your argument is of no legal effect, for where a person injures himself, though it is forbidden, he is nevertheless exempt, yet were others to injure him, they would be liable. So too he who cuts down his own plants, though acting unlawfully, is exempt, yet were others [to do it], they would be liable."[37]

This Mishnah deals with the laws relating to fines. Contrary to the ancient halakhah which laid down that the fine was to be in accordance with the dignity of the person insulted, R. Akiva declared that all are equal in dignity, so that the fine would naturally be the same for all.

Three historical-social circumstances could explain the new statements about the ʿam ha-aretz which first made their appearance in the period of Jabneh.

The first circumstance was the rejection of the sects and streams which had not joined the normative community. It was none other than the abolition of the sects, the demolition of the barriers, and the egalitarian views of the generation of Jabneh, which we have dealt with above, that brought about the need to reject those sects and streams which refused to cooperate in these aims. Although this applied in particular to the Christian and the various Gnostic sects, yet against this background the harsh disapproval of the ʿammei ha-aretz becomes comprehensible, for they too were not prepared to take upon themselves the ways of life and the full obligations imposed upon the people by the Nasi and the Sanhedrin.

The second circumstance was the rise of the class of the Sages. Following the destruction of the Temple, their leadership of the nation comprised an element of political-national rule side by side with their spiritual-religious leadership. Contributing as it did to the advancement of the class of the Sages in the eyes of the nation, this naturally produced a trend towards the crystallization of an exclusive class of Sages, one of the manifestations of which was the creation of a class of the sons of the Sages, attained through זכות אבות "ancestral merit."[38] A con-

[37] Bava Ḳamma viii, 6. בני חרין has the meaning of aristocrats, highly respected persons, nobles (as in Eccl. 10:17 [AV, RV]; and cf. Soṭah ix, 15).

[38] See, for example, TB Berakhot 27b; TJ Berakhot iv, 7d. On the subject of the sons of the Sages, see G. Allon, *Meḥḳarim be-Toledot Yisraʾel*, Tel Aviv, 1957–1958, II, pp. 58–73.

crete illustration of this, though taken from a period after the generation of Jabneh, was the precedence accorded to them when called up to the reading of the Torah.[39]

Echoes of different attitudes to the class of the Sages occur in the following argument. "והעמידו תלמידים הרבה," שבית שמאי אומרים, אל ישנה אדם אלא למי שהוא חכם ועניו ובן אבות ועשיר, ובית הלל אומרים, לכל אדם ישנה, שהרבה פושעים היו בהם בישראל, ונתקרבו לתלמוד תורה, ויצאו מהם צדיקים, חסידים וכשרים "'And raise up many pupils.' Bet Shammai said, One should only teach someone who is wise and meek and of noble descent and rich. But Bet Hillel said, One should teach everyone, for many were the sinners in Israel who were brought near to the study of the Torah, and from them there issued righteous, pious, and honest men."[40] Among the conditions mentioned by Bet Shammai for acceptance by a Yeshivah was also noble descent as a prerequisite for entrance into the world of the Sages.[41] Of the distinctiveness of the Sages as a separate class we learn from the following statement. ...כך תלמידי חכמים ניכרים בהלוכם ובדבורם ובעטפתם בשוק "...So talmidei ḥakhamim are recognized by their walk, by their speech, and by their dress in the street."[42] There are indeed statements dating from various periods, including that of Jabneh, which emphasize the right of everyone to study, as well as the equality of all men before, the Torah. There is, for example, the remark of R. Jose the priest, who was the pupil of Rabban Johanan b. Zakkai. רבי יוסי אומר... והתקן עצמך ללמוד תורה, שאינה ירשה לך "R. Jose said,... And fit yourself to study the Torah for it does not come to you by inheritance."[43] The actual occurrence as also the multiplicity of such statements are to be explained specifically against the background of the rise of the class of the Sages and its development into an exclusive group.

We can thus understand the tension that prevailed between the class of the talmidei ḥakhamim and the stratum of the 'ammei ha-aretz, particularly since matters reached a stage at which privileges were demanded by the talmidei ḥakhamim even if it meant increasing the burden to

---

39 TB Giṭṭin 59b–60a.

40 Avot de-R. Nathan, Version A, iii, ed. Schechter, pp. 14–15.

41 See TB Berakhot 28a: אותו היום סלקוהו לשומר הפתח "On that day the door-keeper was removed," indicating that there was a selective admission to the bet ha-midrash. On the selection of pupils by R. Judah ha-Nasi, see below, p. 186.

42 Sifrei Deuteronomy cccxliii, ed. Finkelstein, p. 400.

43 Avot ii, 12.

be borne by others, such as, for example, exemption from taxes.[44] To this should be added the fact that at all times and in every society a hatred has existed between the intellectual and the uncultured person, a hatred which naturally grows in intensity as the intellectual class advances socially and enjoys special privileges.

The third circumstance, which also had its inception in the period of Jabneh, was the placing of the Torah and its study at the centre of the religious-national life of the people. This circumstance, with which we have previously dealt, had its origin in the leadership and class of the Sages, on the one hand, and in the desire of that leadership, on the other, to fill the vacuum created by the destruction of the Second Temple and the abolition of its ritual.

Accordingly, the severity of the statements against the ʻammei ha-aretz is also to be regarded as being in the nature of propaganda in favour of studying the Torah. It is very probable that one of the means of propagating the study of the Torah and of seeing that every Jew acquired a knowledge of and educated his children in it, was through the severe condemnation and vigorous disapproval of anyone who failed to participate in advancing the study of the Torah.

A new wave of statements against the ʻam ha-aretz, in which harshness reached its peak, occurred in the period of Usha. This can be explained as follows.

In the days after the Bar Kokheva revolt the continued existence of the nation was once more in jeopardy. The results of the Bar Kokheva revolt in the number killed and taken prisoner, in the destruction of settlements and the expropriation of land, were even graver than the situation following the Great Revolt against the Romans. Added to this were the years of religious persecution during which the holding of public assemblies for the study of the Torah or for the teaching of pupils was prohibited,[45] as was also the meeting in synagogues and in batei midrash [houses of study].[46]

After the death of Hadrian and the rise of the Antonines the persecution abated. The Sanhedrin which had been re-established in Galilee

---

[44] TB Bava Batra 8a; etc. Notable evidences of the privileges enjoyed by talmidei ḥakhamim begin to appear in the days of R. Judah ha-Nasi.

[45] TB Pesaḥim 112a; Ta'anit 18a; etc. See M. D. Herr, "Gezerot ha-Shemad ve-Ḳiddush ha-Shem bi-Yemei Hadrianus," *Milḥemet Ḳodesh u-Martyrologyah be-Toledot Yisra'el u-ve-Toledot ha-ʻAmmim*, Historical Society of Israel, Jerusalem, 1967, pp. 73–92.

[46] Tosefta Bava Metziʻa ii, 17; etc. And see Herr, *ibid., loc. cit.*

sought, as had the Sanhedrin in the generation of Jabneh, to refashion the life of the nation and to make the Torah and its study the central and supreme value in the life of the people. In the source testifying to the establishment of the Sanhedrin at Usha, the renewal of the study of the Torah associated with that event is emphasized. בשלפי השמד נתכנסו רבותינו לאושא, ואלו הן, רבי יהודה ורבי נחמיה, רבי מאיר ורבי יוסי ורבי שמעון בן יוחאי ורבי אליעזר בנו של ר׳ יוסי הגלילי ורבי אליעזר בן יעקב, שלחו אצל זקני הגליל ואמרו, כל מי שהוא למד יבוא וילמד, וכל מי שאינו למד יבוא וילמוד. נתכנסו ולמדו ועשו כל צרכיהון "At the end of the period of persecution our Rabbis assembled at Usha. They were R. Judah and R. Nehemiah, R. Meir and R. Jose and R. Simeon b. Yoḥai and R. Eliezer the son of R. Jose the Galilaean and R. Eliezer b. Jacob. They sent to the elders of Galilee saying, Whoever has studied, let him come and teach, and whoever has not studied, let him come and learn. They gathered together and studied and did all that was required of them."[47]

The task which the leadership at Usha undertook of seeking to renew the life of the nation around the Torah and its study was a difficult one. People had presumably refrained from studying the Torah during the years of revolt, during its aftermath, and during the persecution that followed. It was in Galilee that the rehabilitation of the nation and of the leadership took place. While the view of Büchler and of others that Galilee was devoid of the Torah cannot be accepted,[48] it may nevertheless be assumed that, because of its distance from the centre of the country, it had been less occupied in the study of the Torah than Judaea. But it was mainly because of the grave economic crisis which prevailed in the country after the Bar Kokheva revolt that the task of the leadership was made difficult. For at such a time it is obviously hard to induce people to interest themselves in spiritual pursuits.

In this context we can understand the following enactment which dates from the generation of Usha. אמר רב יצחק, באושא התקינו שיהא אדם מתגלגל עם בנו עד שתים עשרה שנה, מכאן ואילך יורד עמו לחייו "R. Isaac said, It was enacted at Usha that a man must bear with his son until he is twelve years of age, after which he may adopt drastic methods."[49] This enactment ordained that a man was obliged to allow his son to pursue his studies until the age of twelve, after which he could be assisted

---

[47] Canticles Rabbah ii, 5.

[48] See p. 7, above, and pp. 200–207, below.

[49] TB Ketubbot 50a.

by his son in earning his livelihood.[50] So grave was the economic position that it was necessary to enact שיהא אדם זן את בניו קטנים "that a man should feed his minor children,"[51] apparently to enable them to obtain an education and to study the Torah.

In the light of this development we may reinterpret the severity of the statements against the 'am ha-aretz in the generation of Usha as having the propaganda aim of reinforcing the study of the Torah by the denunciation, rejection, and shunning of anyone who did not himself acquire a knowledge of the Torah and did not educate his children in it.

It should furthermore be remembered that the period of Usha was one in which the subject of the 'am ha-aretz was dealt with more extensively than at any other time, so that a very large number of harsh statements against the 'am ha-aretz was produced at that period.

Acrimonious expressions against the 'am ha-aretz occurred also under R. Judah ha-Nasi, whose aristocratic, authoritative leadership was a feature of those days.[52] In supporting the process whereby the Sages became a defined, exclusive class, R. Judah ha-Nasi was scrupulous in the selection of pupils, in addition to limiting the freedom of teaching. So, for example, he decreed שלא ישנו לתלמידים בשוק "that pupils were not to be taught in the market place."[53] On good terms with the wealthy and the powerful, with those in the leadership of mixed cities, and others like them, he was friendly disposed to the rich who were themselves 'ammei ha-aretz. Yet matters as a whole were undoubtedly attended by an increasing tension between the Sages and the 'ammei ha-aretz.[54]

It should be added that from a certain point of view the Sages felt themselves more than ever obliged, because of the tension between themselves and the 'ammei ha-aretz, to observe the Torah and standards of moral conduct. So, for example, רבי מאיר שאל את אלישע בן אבויה רבו,

---

50 יורד עמו לחייו is apparently to be interpreted here as referring to a livelihood [כדי חייו]. See *Aruch Completum*, s.v. רד.

51 TJ Ketubbot iv, 28d. Allon, *Toledot ha-Yehudim*, attributes this enactment in one place (II, p. 67) to the centre at Usha after the Bar Kokheva revolt, in another (II, p. 153) to the seat of the leadership at Usha in the days of Rabban Gamaliel b. R. Judah ha-Nasi. The former alternative, namely, that it dates from after the Bar Kokheva revolt, is apparently to be preferred.

52 See, for example, TJ Demai i, 22a; TB Ḥullin 7a–b.

53 TB Mo'ed Ḳaṭan 16a; and see also TB Sanhedrin 5b; TJ Shevi'it vi, 36b–c.

54 In this generation, too, the change began in the attitude of the talmidei ḥakhamim to the 'ammei ha-aretz. On this, see the next section, pp. 188–195.

אמר לו, מהו ״ואשת איש נפש יקרה תצוד,״ אמר לו בן אדם שהוא הדיוט אם נתפש הוא בעבירה אין גנאי לו, למה שהוא אומר הדיוט אני, ולא הייתי יודע עונשה של תורה, אבל חבר אם נתפש בעבירה גנאי הוא לו, מפני שהוא מערב דברי טהרה עם דברי טומאה, אותה תורה שהיתה לו יקרה הוא מבזה אותה, שעמי הארץ אומרים, ראו חבר שנתפש בעבירה ובוזה את תורתו, לכך נאמר ״נפש יקרה תצוד״ "R. Meir asked Elisha b. Avuyah his teacher, What is the meaning of 'But an adulteress stalks a man's very life'? He answered, If a person who is ignorant is caught sinning, it is no disgrace to him. Why? Because he says, I am an ignorant person and did not know the penalty imposed by the Torah. But if a ḥaver[55] is caught sinning, it is a disgrace to him since he confuses pure with impure things, in that he makes the Torah, which was his very life, contemptible, so that the 'ammei ha-aretz say, See, a ḥaver has been caught sinning and spurns his Torah. Therefore it is said, 'It stalks a man's very life'."[56]

Similar demands were made of talmidei ḥakhamim in the following statement. רבי שאל לרבי בצלאל, מהו דכתיב ״כי זנתה אמם,״ אפשר ששרה אמנו זונה היתה, אמר לו חס ושלום, אלא אימתי דברי תורה מתבזין בפני עמי הארץ, בשעה שבעליהן מבזין אותם. אתא רבי יעקב בר אבדימי ועבדיה שמועה, אימתי דברי תורה נעשין כזונות בפני עמי הארץ, בשעה שבעליהן מבזין אותם "Rabbi [R. Judah ha-Nasi] asked R. Bezalel, What is the meaning of the verse, 'For their mother has played the harlot'? Can it be that Sarah our matriarch was a harlot? He answered him, God forbid. But when are the words of the Torah despised by the 'ammei ha-aretz? When talmidei ḥakhamim make them contemptible [through their conduct]. R. Jacob b. Avdimi came and delivered a discourse on it, saying, When do the words of the Torah become like harlots for the 'ammei ha-aretz? When talmidei ḥakhamim make them contemptible."[57]

These demands, it was emphasized, are stricter when made of talmidei ḥakhamim than when made of 'ammei ha-aretz. כדדריש ר׳ יהודה ברבי אלעאי, מאי דכתיב ״והגד לעמי פשעם ולבית יעקב חטאתם״ — ״הגד לעמי פשעם,״ אלו תלמידי חכמים ששגגות נעשות להם כזדונות — ״ולבית יעקב חטאתם,״ אלו עמי הארץ שזדונות נעשות להם כשגגות וכו׳ "As R. Judah b. Ila'i expounded, What is the meaning of 'Declare to my people their transgression, and to the house of Jacob their sins'? 'Declare to my people their transgression,' these are talmidei ḥakhamim, whose unwitting errors are accounted as intentional faults, 'and to the house of Jacob

---

55 Here "ḥaver" has the sense of talmid ḥakham.

56 Midrash Proverbs vi, ed. Buber, p. 56.

57 Ruth Rabbah i, 2.

their sins,' these are the 'ammei ha-aretz, whose intentional sins are accounted to them as unwitting errors..."[58]

3. The Change in the Attitude of the Sages to the 'Ammei ha-Aretz

Beginning in the days of R. Judah ha-Nasi and continuing in those of the first generations of the Eretz Israel Amoraim, the Sages modified their attitude to the 'am ha-aretz la-Torah, even to the extent of vindicating the 'ammei ha-aretz despite their want of knowledge of the Torah.

The following incident to some extent mirrors the process of the change that took place. רבי פתח אוצרות בשני בצורת, אמר יכנסו בעלי מקרא, בעלי משנה, בעלי גמרא, בעלי הלכה, בעלי הגדה, אבל עמי הארץ אל יכנסו, דחק רבי יונתן בן עמרם ונכנס, אמר לו רבי פרנסני, אמר לו, בני קרית, אמר לו לאו. שנית, אמר לו לאו. אם כן במה אפרנסך. אמר לו פרנסני ככלב וכעורב, פרנסיה. בתר דנפק יתיב רבי וקא מצטער, ואמר אוי לי שנתתי פתי לעם הארץ, אמר לפניו רבי שמעון בר רבי, שמא יונתן בן עמרם תלמידך הוא, שאינו רוצה ליהנות מכבוד תורה מימיו. בדקו ואשכח, אמר רבי יכנסו הכל "Rabbi [R. Judah ha-Nasi] once opened his storehouse [of victuals] in a year of scarcity, proclaiming, Let those enter who have studied Scripture, or Mishnah, or gemara, or the halakhah, or the haggadah, but let not the 'ammei ha-aretz enter. R. Jonathan b. Amram pushed his way in and said, Master, give me food. He said to him, My son, have you learnt Scripture? He replied, No. Have you learnt Mishnah? No. If so, he said, how can I give you food? He said, Feed me as you would a dog and a raven. He gave him some food. After he had left, R. Judah ha-Nasi's conscience smote him and he said, Woe is me that I have given my bread to an 'am ha-aretz. R. Simeon, the son of R. Judah ha-Nasi, said to him, Perhaps it is Jonathan b. Amram, your pupil, who has all his life made it a principle not to derive material benefit from the honour paid to the Torah. Inquiries were made and it was found that it was so, whereupon R. Judah ha-Nasi said, All may now enter."[59] Two trends, contrary to those discussed in the previous section, find expression in this incident, the one, the dissatisfaction of the talmidei ḥakhamim themselves with the privileges of their class, the other, the first signs of a

---

[58] TB Bava Metzi'a 33b.

[59] TB Bava Batra 8a; and cf. Kallah Rabbati ii, 14, ed. Higger, p. 209.

change in the attitude to the 'ammei ha-aretz. A tradition about a change in R. Judah ha-Nasi's attitude to the 'ammei ha-aretz, as compared to that prelvaent in the generation of Usha, occurs in the following statement. רבי סבר ניחא ליה לחבר דלעביד הוא איסורא קלילא, ולא ליעבד עם הארץ איסורא רבה, ורבן שמעון בן גמליאל סבר ניחא ליה לחבר דליעבד עם הארץ איסורא רבה, ואיהו אפילו איסורא קלילא לא ליעבד "Rabbi [R. Judah ha-Nasi] held that a ḥaver[60] is satisfied to commit a minor ritual offence in order that an 'am ha-aretz should not commit a major one, while Rabban Simeon b. Gamaliel held that a ḥaver prefers that an 'am ha-aretz commits a major offence rather than that he himself should commit even a minor one."[61]

In these generations the Sages arrived at the conclusion that a healthy nation is composed of various elements which complement and guarantee the survival of one another, and that the unity of the nation depends upon proper relations between these elements.[62] To reinforce this view, several sources compared the nation to a vine. אמר רבי שמעון בן לקיש, אומה זו כגפן נמשלה, זמורות שבה אלו בעלי בתים, אשכולות שבה אלו תלמידי חכמים, עלין שבה אלו עמי הארץ, קנוקנות שבה אלו ריקנים שבישראל, והיינו דשלחו מתם, ליבעי רחמים איתכליא על עליא, דאילמלא עליא, לא מתקיימין איתכליא "R. Simeon b. Laḳish said, This people [Israel] is like a vine, its branches are the aristocracy, its clusters the talmidei ḥakhamim, its leaves the 'ammei ha-aretz, its twigs those in Israel who are empty of learning. This is what was meant when they sent from there [Eretz Israel], saying, Let the clusters pray for the leaves, for were it not for the leaves, the clusters could not exist."[63] Similes of this kind also emphasized that a person, even though an 'am ha-aretz, could nevertheless have good qualities which could confer privileges on him. "פרי עץ הדר" אלו ישראל, מה אתרוג זה יש בו טעם ויש בו ריח, כך ישראל יש בהם בני אדם שיש בהם תורה ויש בהם מעשים טובים. "כפות תמרים" אלו ישראל, מה התמרה הזו יש בה טעם ואין בה ריח, כך הם ישראל, יש בהם בני אדם, שיש בהם תורה, ואין בהם מעשים טובים. "וענף עץ עבות" אלו ישראל, מה הדס זה יש בו ריח ואין

---

60 Here used in the sense of talmid ḥakham.

61 TB 'Eruvin 32b, where the reference is to tithes; and cf. TJ Ma'aserot ii, 49c.

62 See Urbach, *Ḥazal*, pp. 477, 580–584.

63 TB Ḥullin 92a; and see Diḳduḳei Soferim. It is interesting to note that Resh Laḳish also said עם הארץ אני אצל הטהרות "I am an 'am ha-aretz as far as purities are concerned" (TJ Demai ii, 23a). Cf. Leviticus Rabbah xxxvi, 2, ed. Margulies, pp. 838–839: מה הגפן הזאת העלים שלה מכסים אשכלותיה, כך הן ישראל עמי הארץ מכסין תלמידי חכמים "Even as the leaves on this vine protect its bunches of grapes, so in Israel the 'ammei ha-aretz protect the talmidei ḥakhamim."

בו טעם, כך הם ישראל יש בהם בני אדם שיש בהם מעשים טובים ואין בהם תורה. "וערבי נחל" אלו ישראל, מה ערבה זו אין בה לא טעם ולא ריח, כך הן ישראל, יש בהן בני אדם, שאין בהן לא תורה ולא מעשים טובים. ומה הקדוש ברוך הוא עושה להן, לאבדן אי איפשר, אלא אמר הקדוש ברוך הוא, יוקשרו כולן אגודה אחת, והן מכפרים אילו על אילו " 'The fruit of goodly trees': these refer to Israel. As a citron has taste and smell, so in Israel some have both Torah and good deeds. 'Branches of palm-trees': these refer to Israel. As the date has taste but no smell, so in Israel some have Torah but no good deeds. 'And boughs of leafy trees': these refer to Israel. As the myrtle has smell but no taste, so in Israel some have good deeds but no Torah. 'Willows of the brook': these refer to Israel. As the willow has neither taste nor smell, so in Israel some have neither Torah nor good deeds. What does the Holy One, blessed be he, do with them? To destroy them is impossible. Therefore the Holy One, blessed be he, said, Let them all be bound together in one bundle and they will atone for one another."[64]

R. Yannai, one of the most important of the first generation of Eretz Israel Amoraim, sought to show that an ʻam ha-aretz could have qualities possessed not even by Sages. The account of this has a literary theme reminiscent of the one which tells that R. Judah ha-Nasi "once opened his storehouse [of victuals] in a year of scarcity." מעשה ברבי ינייי שהיה מהלך בדרך, פגע בו אדם אחד שהיה משופע ביותר, אמר לו משגח רבי לאיתקבלא גבן, אמר ליה מה דהני לך, הכניסו לתוך ביתו. בדקו במקרא ולא מצאו, בדקו במשנה ולא מצאו, בתלמוד ולא מצאו, בהגדה ולא מצאו, אמר ליה, בריך, אמר ליה יברך ינייי בבייתיה. אמר ליה אית בך אמר מה דאנא אמר לך, אמר ליה אין. אמר ליה אמור אכל כלבא פסתיה דייני, קם צריה, אמר ליה מה ירתותי גבך דאת מוני לי, אמר ליה ומה ירתותך גבי, אמר ליה דמינוקייה אמ׳ "תורה צוה לנו משה מורשה," קהלת ינייי אין כת׳ כאן, אלא "קהלת יעקב." מן דאיתפיסין דין לדין אמר ליה למה זכית למיכל על פתורי, אמר ליה מן יומיי לא שמעית מילא בישא וחיזרתי למריה, ולא חמית תרין דמיתכתשין דין עם דין ולא יהבית שלמא ביניהון. אמר ליה כל הדא דרך ארץ גבך וקריתך כלבא, וקרא עליה "ושם דרך אראנו בישע אלהים" "It once happened that R. Yannai, when walking along the road, met a man who was well dressed [giving the impression of being a talmid ḥakham] and to whom he said, Will it please the Rabbi to be our guest? He answered, As it pleases you. R. Yannai took him to his home and tested him in Scripture but found he had none, tested him in Mishnah but found he had none, in Talmud but found he had none, in haggadah but found he had none. He said to him, Say the blessing [for the grace

---

[64] Leviticus Rabbah xxx, 19, ed. Margulies, p. 709; and parallel passages.

after meals]. He answered, Let Yannai say the blessing in his own home. He said, Can you repeat what I say to you? He answered, Yes. He said, Say, a dog has eaten Yannai's bread. The man stood up and shouting at him, said, You have my heirloom which you withhold from me. And what heirloom of yours, asked Yannai, have I? He answered, I have heard children say, 'Moses commanded us a Torah, as a possession for the assembly of Jacob.' The verse does not say, 'for the assembly of Yannai' but 'for the assembly of Jacob.' After they had asked each other's pardon, Yannai said to him, How is it that you have had the merit to eat at my table? He answered, I have never heard an unkind word and reported it to the one about whom it was said, and I have never seen two people quarrelling without making peace between them. Thereupon Yannai said to him, You have so much derekh eretz [good manners and breeding] and I called you a dog. And he applied to him the verse, 'To him that orders his way [derekh] aright I will show the salvation of God'."[65]

Sages in these generations sought to find a remedy for the 'ammei ha-aretz, as can be seen, for example, from the discussion between the Amora R. Eleazar and R. Johanan. אמר רבי אלעזר, עמי הארצות אינן חיים, שנאמר "מתים בל יחיו"... כיון דחזייה [את רבי יוחנן] דקמצטער, אמר ליה, רבי מצאתי להן תקנה מן התורה, "ואתם הדבקים בה׳ אלהיכם חיים כולכם היום," וכי אפשר לדבוקי בשכינה, והכתיב "כי ה׳ אלהיך אש אוכלה," אלא כל המשיא בתו לתלמיד חכם, והעושה פרקמטיא לתלמידי חכמים והמהנה תלמידי חכמים מנכסיו, מעלה עליו הכתוב כאילו מדבק בשכינה "R. Eleazar said, The 'ammei ha-aretz will not be resurrected, as it is said, 'They are dead, they will not live.' ...When he saw that he [R. Johanan] was distressed, he said to him, Master, I have found a remedy for them from the Torah, 'But you who held fast to the Lord your God are all alive this day.' Now is it possible to hold fast to the divine presence, of whom it is written, 'For the Lord your God is a devouring fire'? But [the meaning is] whoever marries his daughter to a talmid ḥakham, or carries on trade on behalf of talmidei ḥakhamim, or benefits talmidei ḥakhamim from his estate is regarded by Scripture as though he has held fast to the divine presence."[66] What is also notable in this source is the transition from the negative approach to the 'am ha-aretz, which goes so far as to proclaim that he has no share in the resurrection, to a positive atti-

---

[65] Leviticus Rabbah ix, 3, ed. Margulies, pp. 176–178. An incident with a similar theme is also told about R. Simeon b. Eleazar (TB Ta'anit 20a–b).

[66] TB Ketubbot 111b.

tude to him, which is however conditional on his establishing close relations with and on his supporting talmidei ḥakhamim. It is interesting to note that one of the pieces of advice given to the 'am ha-aretz is that he should marry his daughter to a talmid ḥakham, as if all that had previously been said about the marriage of a talmid ḥakham to the daughter of an 'am ha-aretz had been forgotten.

An end to the hatred of the 'am ha-aretz is explicitly advocated in the following statement which seeks to explain the words of R. Joshua b. Hananiah who included שנאת הבריות "hatred of one's fellowmen"[67] among the things that מוציאין את האדם מן העולם "take a man out of the world": ושנאת הבריות כיצד, מלמד שלא יכוין אדם לומר, אהוב את החכמים ושנא את התלמידים, אהוב את התלמידים ושנא את עמי הארץ, אלא אהוב את כולם, ושנא את המינין ואת המשומדים ואת המסורות וכו׳ "What is hatred of one's fellowmen? It teaches that one should not intend to say, Love the Sages and hate the pupils, love the pupils and hate the 'ammei ha-aretz, but love all of them and hate the heretics, the apostates, and the informers..."[68]

Contrary to the halakhot and the statements which demand of talmidei ḥakhamim that they shun the homes and places of assembly of the 'ammei ha-aretz, the talmidei ḥakhamim were now enjoined to teach the Torah to the 'ammei ha-aretz. ״ועניים מרודים תביא בית,״ אילו תלמידי חכמים שהן נכנסין בבתיהן של עמי הארץ, ומרווין אותן מדברי תורה "'And bring the homeless poor into your house,' these refer to the talmidei ḥakhamim who enter the homes of the 'ammei ha-aretz and permeate them with the words of the Torah."[69]

The apex of close relations between the 'ammei ha-aretz and the talmidei ḥakhamim is to be found in the following incident which tells of a remark that was made by an 'am ha-aretz, that was accepted by a Sage, and was expounded in the former's name in the bet ha-midrash. חד עם דארע אמר ליה לרבי הושעיא, אין אמרית לך חדא מילה טבא את אמ׳ לה משמי בציבורא, אמר ליה מה היא, אמר ליה כל הדוריות שנתן אבינו יעקב לעשו עתידין אומות העולם להחזירן למלך המשיח לעתיד לבוא, מה טעם, ״מלכי תרשיש ואיים מנחה,״ יביאו אין כתיב, אלא ״ישיבו.״ אמר ליה, חייך מילה טבא אמרת, משמך אנא אמ׳ לה "An 'am ha-aretz said to R. Hoshaia,[70] If I tell you something good, will you repeat it in public in my name? He asked,

[67] Avot ii, 11.

[68] Avot de-R. Nathan, Version A, xvi, ed. Schechter, p. 64.

[69] Leviticus Rabbah xxxiv, 13, ed. Margulies, p. 801.

[70] An Eretz Israel Amora of the first half of the third century CE.

What is it? He answered, All the gifts which our patriarch Jacob gave to Esau, the nations of the world will in the messianic future return to the king Messiah. Why? [Because it says], 'The kings of Tarshish and the isles shall render [lit. "restore"] tribute.' Scripture does not say 'shall bring' but 'shall restore.' He [R. Hoshaia] said to him, By your life, it is a good remark that you have made. I shall quote it in your name."[71]

The reasons for the change in the attitude to the 'ammei ha-aretz are as follows.

Mention should first be made of two social factors in the days of R. Judah ha-Nasi. The one was R. Judah ha-Nasi's general desire, expressed in his enactments and halakhot, to normalize Jewish life in Eretz Israel. Social normalization was obviously impossible against a background of tension and hatred between the classes of the talmidei ḥakhamim and the 'ammei ha-aretz. The other factor was the close relations established by R. Judah ha-Nasi with the rich and the powerful, who were themselves 'ammei ha-aretz. Thus, for example, we find the Sages mocking and sneering at R. Judah ha-Nasi's son-in-law, Ben Elasah, on account of his ignorance of the Torah.[72] R. Judah ha-Nasi could not have it both ways: on the one hand, מכבד עשירים "honouring the rich"[73] who were far removed from the Torah and, on the other, advocating that the 'ammei ha-aretz in the form of the masses be shunned.

After the days of R. Judah ha-Nasi the office of the Nasi was separated from the Sanhedrin, with the national-political leadership in the hands of the former and the spiritual-religious leadership in those of the latter. This separation deprived the Sanhedrin and the Sages of the basis of their rule and of their crystallization into a distinct, exclusive class, which was one of the reasons for the tension that existed between them and the 'ammei ha-aretz. It now became possible for the Sages to maintain their popular character, and their relations with the mass of the people were re-established. Against this background we can explain the

---

71 Genesis Rabbah lxxviii, 12, ed. Theodor-Albeck, pp. 932–933. Here we have an illustration of the way in which talmudic literature was composed, comprising as it does not only a collection of the statements of Sages but also the additions of pupils and non-pupils. And see S. Safrai, "Elementary Education, Its Religious and Social Significance in the Talmudic Period," *Cahiers d'Histoire Mondiale*, 11 (1968), pp. 148–169.

72 TJ Mo'ed Ḳaṭan iii, 81c.

73 TB 'Eruvin 86a.

reduced tension between the Sages and the ʻammei ha-aretz, the former's closer relations with the latter, and the Sages' recognition that good qualities and moral values could be possessed by those with a sparse knowledge of the Torah.

In earlier chapters we dealt with the obscurity that marked the concept of the ʻam ha-aretz le-mitzvot at the beginning of the period of the Eretz Israel Amoraim. We said that, as regards their actual social components, no distinction existed between the stratum which may be referred to as the ʻammei ha-aretz le-mitzvot and that which may be called the ʻammei ha-aretz la-Torah. This undoubtedly contributed to reducing the significance of the concept of the ʻam ha-aretz la-Torah and naturally conduced to bridging the gulf between the ʻam ha-aretz la-Torah and the Sages.

The main impetus to the closer relations between the Sages and the ʻammei ha-aretz undoubtedly came from the historical conditions prevailing in the period of anarchy which occurred in the third century CE. A considerable number of the statements which have been cited contain the demand that the ʻammei ha-aretz support the talmidei ḥakhamim, a demand that is to be understood against the background of the grave economic crisis which occurred in the period of anarchy. This crisis is not, however, to be compared to that which followed the Bar Kokheva revolt. Whereas the earlier crisis which followed the Bar Kokheva revolt and which witnessed increased tension between the talmidei ḥakhamim and the ʻammei ha-aretz came about through internal events and was accompanied by religious persecution, the later crisis of the third century anarchical period, which was even graver than that following the Bar Kokheva revolt, was caused by external factors arising from the instability of the imperial regime. Now no need arose, as there had after that earlier crisis, to restore national life to its former state, since neither the leadership of the nation nor its relations with the authorities had been undermined. Instead there was now the need literally to ensure the physical survival of the nation. The study of the Torah and the continued existence of the class of the talmidei ḥakhamim were, but for the help of the people and the support of the ʻammei ha-aretz, in danger of extinction.

The general feeling of the nation in the period of anarchy in the third century CE was one of shared distress. In the nature of things this feeling reinforced the unity of the different sections of the nation. Not only did the harsh expressions against the ʻammei ha-aretz cease in this generation, not only was there a certain "going down" by the

Sages to the people, but the people also sensed this closer relation with the leadership, for כשישראל בצרה, אף גדוליהם מצטערים עמהם "when Israel is in trouble, even their leaders grieve with them."[74] [75]

4. The Sons of the 'Ammei ha-Aretz

The emergence, after the destruction of the Temple, of the Sages as an exclusive class had several consequences. It led in the course of time to the demand by the Sages for privileges for themselves. It constituted, as we have seen, one of the causes of the hatred between them and the 'ammei ha-aretz. And it gave rise, on the one hand, to a class of בני חכמים "the sons of the Sages" and, on the other, to that of בני עמי הארץ "the sons of the 'ammei ha-aretz." The former expression refers to those who enjoyed privileges by virtue of being the sons of Sages, without being themselves in every instance also learned in the Torah,[76] while the latter term included both those who were themselves 'ammei ha-aretz and those who, becoming attached to the Torah, sought to join the world of the Torah and the circle of the Sages.

In every generation we come across Sages who had no personal distinction and no "ancestral merit" [zekhut avot], whose families were undistinguished, and some of whom were even the offspring of proselytes. The most extreme case is that of R. Akiva who, it is emphasized, was himself in his early years an 'am ha-aretz and without "ancestral merit." This may also be the reason for R. Akiva's many statements stressing that the Torah belongs equally to all, and that all have an equal right to study it. He also expressed the equality of all Israel, in quite another context, by declaring that כל ישראל ראוין לאותה איצטלא "all Israel are worthy of that [princely] robe."[77] R. Akiva who, when

---

[74] Pesiḳta Rabbati xii, ed. Friedmann, 50b.

[75] It should be mentioned that there remained an indirect opposition to the 'am ha-aretz in one area, and that was in the appointment of dayyanim [judges] who were 'ammei ha-aretz and ignorant. Here the opposition was expressed in particular against those who appointed such dayyanim. See above, pp. 106–107; and see also Allon, *Meḥḳarim*, II, pp. 15–57.

[76] See above, pp. 182–183; and see Allon, *Meḥḳarim*, I, pp. 58–73; *idem*, *Toledot ha-Yehudim*, I, pp. 315–318; E. E. Urbach, "Ma'amad ve-Hanhagah be-'Olamam shel Ḥakhmei Eretz Yisra'el," *Divrei ha-Aḳademyah ha-Le'ummit ha-Yisra'elit le-Madda'im*, II, 4, Jerusalem, 1969, pp. 31–54; *idem*, *Ḥazal*, pp. 465–584.

[77] TB Bava Metzi'a 113b; and cf. the remark of his outstanding pupil, R. Simeon b. Yoḥai: כל ישראל בני מלכים הם "All Israel are princes" (TB Shabbat 67a, 128a; and cf. Bava Metzi'a 113b).

an ʿam ha-aretz, hated, as we have seen, talmidei ḥakhamim was well aware of the reasons for that hatred, and it is against this background that his remarks against the pride and conceit of talmidei ḥakhamim are to be understood. אמר רבי עקיבא, כל המגביה עצמו על דברי תורה למה הוא דומה, לנבלה מושלכת בדרך, כל עובר ושב מניח ידו על חוטמו ומתרחק ממנה והולך וכו׳ "R. Akiva said, Whoever exalts himself above the words of the Torah is like a carcass lying exposed on the road. Every passer-by holds his nose and gets as far away from it as he can..."[78] or רבי עקיבה אומר, מי גרם לך להתנבל בדברי תורה, על ידי שנשאת עצמך בהם "R. Akiva said, What was the cause of your disgrace [by exposing your ignorance] in the words of the Torah? It is because you exalted yourself through them."[79]

Although R. Akiva became one of the most important Sages, if not the most important Sage, of his generation, he was nonetheless not appointed Nasi when Rabban Gamaliel was temporarily deposed. The reason mentioned in the sources is that he did not have "ancestral merit,"[80] and R. Akiva himself deplored this. והיה רבי עקיבה יושב ומצטער ואמר לא שהוא [רבי אלעזר בן עזריה שהתמנה במקום רבן גמליאל] בן תורה יותר ממני, אלא שהוא בן גדולים יותר ממני, אשרי אדם שזכו לו אבתיו, אשרי אדם שיש לו יתד במי להתלות בה "And R. Akiva sat and was distressed and said, It is not that he [R. Eleazar b. Azariah who was appointed to replace Rabban Gamaliel] is more learned in the Torah than I am, but that he is the son of more distinguished progenitors. Happy is the man whose ancestors have transmitted privileges to him; happy is the man who has a peg to hang on."[81] It was apparently this background that engendered R. Akiva's opposition to the rule of talmidei ḥakhamim, based as it was on pedigree and on wealth, and hence his advice to his son, אל תדור בעיר שראשיה תלמידי חכמים "Do not dwell in a city whose leaders are talmidei ḥakhamim."[82]

The history of R. Akiva shows that it was not easy for an ʿam ha-aretz to enter the circle of the Sages, and even if he succeeded in doing so, he still suffered from the fact that he had no distinguished ancestry.

78 Avot de-R. Nathan, Version A, xi, ed. Schechter, p. 46.

79 Genesis Rabbah lxxxi, 2, ed. Theodor-Albeck, p. 969.

80 TB Berakhot 27b. Already the Babylonian Talmud, unable to grasp the significance of "ancestral merit" as a prerequisite for the position of Nasi, explains דילמא עניש ליה דלית ליה זכות אבות "Perhaps he [Rabban Gamaliel] will bring a curse on him because he has no ancestral merit."

81 TJ Berakhot iv, 7d.

82 TB Pesaḥim 112a.

As a rule the 'ammei ha-aretz and the sons of the 'ammei ha-aretz did not establish close relations with the circle of the Sages, and statements against matrimonial ties between the families of Sages and those of the 'ammei ha-aretz bear witness to this. All the remarks, after the manner of those of R. Akiva, against the pride and conceit of the Sages and the sons of the Sages also indicate that the Sages shunned and segregated themselves from the 'ammei ha-aretz. This naturally led to the estrangement of the 'ammei ha-aretz and of the sons of the 'ammei ha-aretz from the Torah and from its study.[83]

The uncertainty whether the son of an 'am ha-aretz could join the circle of the Sages and enter the world of the Torah was felt by the sons of the 'ammei ha-aretz themselves. "מורשה" ולא ירושה, שלא יאמר בן עם הארץ הואיל ואין אבי תלמיד חכם מה אני מועיל אם אלמד תורה, אלא "יזל מים מדליו," מן הדלים שבו "'Possession' and not 'inheritance,' that the son of an 'am ha-aretz should not say, Since my father is not a talmid ḥakham, what use is it if I study the Torah, but 'water shall flow from his buckets [mi-dalyav],' from the poor [dallim] among them."[84]

---

83 Analogous to the question of the right of the sons of the 'ammei ha-aretz to engage in the study of the Torah was the problem of the right of the proselyte to do so. It was undoubtedly against the background of opposition to the proselyte's study of the Torah that the following halakhah was laid down. וראה גר שבא ללמוד תורה, לא יאמר, ראו מי בא ללמוד תורה, שאכל נבילות וטריפות שקצים ורמשים, וכן הוא אומר "ויען איש משם ויאמר מי אביהם," וכי יש אב לתורה, והלא כבר נאמר "מה שמו ומה שם בנו כי תדע," ואומר "בית והון נחלת אבות ומה' אשה משכלת" "And [if] he sees a proselyte coming to study the Torah, he should not say, 'See who comes to study the Torah, one who ate nevelot [animals not slaughtered according to Jewish ritual law] and ṭerefot [animals with an organic disease] and all kinds of forbidden food.' And thus it is said, 'And a man of the place answered, "And who is their father?"' Has the Torah then a father? Has it not been said, 'What is his name, and what is his son's name? Surely you know!' And it is also said, 'House and wealth are inherited from fathers, but a prudent wife is from the Lord'" (Tosefta Bava Metzi'a iii, 25; and cf. also Midrash Psalms, Ps. 1, xviii, ed. Buber, p. 18; Numbers Rabbah xiii, 16: in these Midrashim the proselyte who studies the Torah is compared to a high priest. This was undoubtedly intended as a protest against those who opposed the proselytes' joining the class of the Sages and engaging in the study of the Torah).

84 Midrash Tannaim to Deut. 33:4, pp. 212–213. Similar to this is the statement: שמא תאמר, ישנו בני הזקנים, ישנו בני הגדולים, ישנו בני הנביאים, תלמוד לומר "כי אם שמור תשמרון" – מגיד שהכל שוים בתורה. וכן הוא אומר "תורה ציוה לנו משה מורשה קהלת יעקב" – כהנים לוים וישראלים אין כתוב כאן, אלא "קהלת יעקב" וכו' "Perhaps you will say, There are the sons of the elders, there are the sons of the great, there are the sons of the prophets. Therefore the text says, 'If you will be careful to do,' thus declaring that all are equal in the Torah. And so too it says, 'Moses

Corresponding to the change in the Sages' attitude to the ʿammei ha-aretz, mentioned above, there also came a change in the attitude to the sons of the ʿammei ha-aretz. אמר רב חמא, מאי דכתיב ״בלב נבון תנוח חכמה ובקרב כסילים תודע,״ בלב נבון תנוח חכמה, זה תלמיד חכם בן תלמיד חכם, ובקרב כסילים תודע, זה תלמיד חכם בן עם הארץ "R. Ḥama said, What is meant by the verse 'Wisdom abides in the mind of a man of understanding, but it is known in the heart of fools'? 'Wisdom abides in the mind of a man of understanding,' this refers to a talmid ḥakham the son of a talmid ḥakham. 'But it is known in the heart of fools,' this refers to a talmid ḥakham the son of an ʿam ha-aretz."[85] In pursuance of this tendency the Sages sought to intensify the teaching of the Torah to the sons of the ʿammei ha-aretz. אמר רבי יונתן, כל המלמד את בן חבירו תורה זוכה ויושב בישיבה של מעלה, שנאמר ״אם תשוב ואשיבך לפני תעמוד,״ וכל המלמד את בן עם הארץ תורה, אפילו הקדוש ברוך הוא גוזר גזירה מבטלה בשבילו, שנאמר ״ואם תוציא יקר מזולל כפי תהיה״ "R. Jonathan said, Whoever teaches the Torah to the son of his fellowman is privileged to sit in the heavenly academy, for it is said, 'If you return, I will restore you, and you shall stand before me.' And he who teaches the Torah to the son of an ʿam ha-aretz, even if the Holy One, blessed be he, has issued a decree he annuls it for his sake, as it is said, 'If you utter what is precious, and not what is worthless, you shall be as my mouth'."[86] In several parallel passages there is the statement והזהרו בבני עמי הארץ שמהן תצא תורה "And take heed of the sons of the ʿammei ha-aretz, for from them shall go forth the Torah."[87] This statement was clearly made against the background of the outlook with which we have previously dealt and which sought to limit the Torah and its study to the class of the Sages and the sons of the Sages.

---

commanded us a Torah, as a possession for the assembly of Jacob.' Priests, Levites, Israelites are not mentioned here, but 'the assembly of Jacob'..." (Sifrei Deuteronomy xlviii, ed. Finkelstein, p. 112).

[85] TB Bava Metziʿa 85b.

[86] TB Bava Metziʿa 85a.

[87] TB Sanhedrin 96a, in the name of R. Judah b. Bathyra of Nisibis, but it is doubtful whether this tradition is to be ascribed to him. See Büchler, *Der galiläische ʿAm-haʾAreṣ*, pp. 19–20, note 3; Allon, *Meḥḳarim*, I, p. 63, note 19. Cf. also חמישה דברים שלח רבי יהודה מנציבין... והיזהרו [בבני] עמי הארץ, שמהם תצא תורה [לישראל] "R. Judah of Nisibis sent five things... And take heed [of the sons of] the ʿammei ha-aretz, for from them shall go forth the Torah [to Israel]" (Pirḳa de-Rabbenu ha-Ḳadosh, Bava de-Ḥamishah, ed. Schönblum, 20b; and see A. Jellinek, *Bet ha-Midrasch*, ii, p. 95; and see M. Higger, "Pirḳei Rabbenu ha-Ḳadosh," *Ḥorev*, 6 [1941], pp. 116–149).

Against the trend that endeavoured to make it impossible for the sons of the ʻammei ha-aretz to enter the circle of the Sages, it was emphasized that generally the Torah was not to be found among the sons of talmidei ḥakhamim. ומפני מה אין מצויין תלמידי חכמים לצאת תלמידי חכמים מבניהן, אמר רב יוסף, שלא יאמרו תורה ירושה היא להם "And why is it not usual for talmidei ḥakhamim to have sons who are talmidei ḥakhamim? R. Joseph said, That people should not say that the Torah comes to them by inheritance."[88]

The halakhah does not deal much with the sons of ʻammei ha-aretz who may have wished to remain ʻammei ha-aretz, or may have expressed no desire, or none was expressed on their behalf, to study the Torah. Evidence of sons who were ʻammei ha-aretz and were quite content to remain as such can be found in the following statement. משל לאב בית דין שהיו לו בנים הרבה, והיו כולן עמי הארץ, ובנו הגדול בן תורה. כתב לכל אחד ואחד חלקו חוץ מבנו הגדול. אמר, לכל אחד נתת חלק, ולי אין אתה כותב, אמר לו דייך תפוש את מקומי "It is like an av bet din [the president of the court] who had many sons, all of whom were ʻammei ha-aretz save the eldest son who was learned in the Torah. To each he [the father] assigned his portion in writing except to his eldest son, who said, To each you have given a portion but to me you have assigned nothing, whereupon he answered him, It is enough that you will take my place."[89]

This statement comprises several noteworthy features. i) The position of av bet din was bequeathed by the holder of the office to his son, thus assuring him of a livelihood. ii) It was apparently not by chance that the position was bequeathed to the eldest son, the only one who was learned in the Torah and who was presumably prepared for it from the outset. iii) This source attests that the sons who were ʻammei ha-aretz and ignorant of the Torah and without any desire to be learned in it recognized that it was specifically they who received the inheritance. If this statement may be attributed to a particular period, Urbach is apparently correct in assigning it to the days of R. Judah ha-Nasi, for it was then that the change in the attitude to the ʻam ha-aretz took place and close relations were established between the Sages and the rich.[90]

---

[88] TB Nedarim 81a. Cf. the remark of R. Jose, the pupil of Rabban Johanan b. Zakkai, which we have cited above: והתקן עצמך ללמוד תורה שאינה ירשה לך "And fit yourself to study the Torah for it does not come to you by inheritance" (Avot ii, 12).

[89] Yalḳuṭ Shimʻoni on Deut. 18:2, §915.

[90] See Urbach, *Divrei ha-Aḳademyah ha-Le'ummit ha-Yisra'elit le-Maddaʻim*, II, 4, 1969, p. 52.

CHAPTER SIX

THE 'AMMEI HA-ARETZ IN JUDAEA AND IN GALILEE

We have previously dealt with the view, held by some scholars, that Galilee was inhabited by the 'ammei ha-aretz.[1] These scholars maintained that in the days of the Second Temple and in the generation of Jabneh the Jewish character of Galilee was actually in doubt, seeing that it was populated by 'ammei ha-aretz who were devoid of the Torah and remote from Judaism. From this world there emerged Galilaean Christianity. To the Galilaean 'ammei ha-aretz came Jesus and his disciples and fought the battle of the 'ammei ha-aretz against the Pharisees and the Sages.[2] It was, so these scholars held, only after the Bar Kokheva revolt that the Jewish character of Galilee was renewed, when many Judaeans under the leadership of the Nasi and the Sanhedrin migrated to Galilee and there propagated anew the spirit of the Torah and of the halakhah.

This view, as advanced by most if not all the scholars who held it, was marked by an apologetic element.

In the course of extolling the activities of Jesus, Protestant scholars of the last century, as well as others who followed in their footsteps, dealt with the subject of the 'am ha-aretz, in that they maintained that Jesus led a class and social reform aimed at raising the status of the 'ammei ha-aretz — the ordinary people — who populated Galilee and whom the spiritual leadership of the nation despised and rejected.

Jewish scholars who held this view did so in a repudiation of Christianity and a denunciation of the 'ammei ha-aretz whom they saw as lacking national sentiment and as concerned only with themselves and with their personal advantage.[3]

Büchler, who forced the concept of the 'am ha-aretz le-mitzvot into the days following the Bar Kokheva revolt, also, in fact, agreed with the view that there was a lack of a knowledge of the Torah and of an observance of the commandments in Galilee in the days of the Second

---

[1] See above, pp. 2, 4–7; see also p. 2, note 6, and p. 4, notes 15 and 16.

[2] See also below, pp. 218–220.

[3] See, for example, J. Klausner, *Hisṭoryah shel ha-Bayit ha-Sheni*, Jerusalem, 1958^5, IV, pp. 220–223.

Temple and in the generation of Jabneh. According to him, the concept of the ʻam ha-aretz le-mitzvot came into being against the background of the attempt by the Sages of Usha, who arrived in Galilee after the Bar Kokheva revolt, to spread the observance of the commandments among the people.[4] The forcing of the concept of the ʻam ha-aretz le-mitzvot into the Procrustean bed of this period had its roots in apologetics directed against the view of the above-mentioned Protestant scholars who associated the concept of the ʻam ha-aretz with the emergence of Christianity. Although seeking to dissociate the two, fundamentally Büchler nevertheless shared the view which regards Galilee in the period under consideration as empty of the Torah and devoid of the observance of the commandments.

Some contemporary scholars, too, connect the existence of the ʻammei ha-aretz in particular with Galilee. Thus, for example, Baron states quite definitely that Galilee was the main centre of the ʻammei ha-aretz.[5]

The scholars who deny any Jewish character to Galilee base themselves on several evidences taken from both non-talmudic and talmudic literature. Of these evidences the main ones are as follows.

The New Testament mentions several times that Pharisees and Scribes came from Jerusalem to Galilee. So, for example, "... the Pharisees gathered together to him [to Jesus after his arrival at Gennesaret], with some of the Scribes, who had come from Jerusalem."[6] From this, various scholars conclude that since the Pharisees and Scribes came from Jerusalem to Galilee, their purpose was to rouse the people of Galilee from their ignorance and bring them under the yoke of the Pharisaic Torah, and hence it follows that there were no such Sages in Galilee itself.[7]

Josephus in his account of the founding of Tiberias tells that Herod Antipas, in building Tiberias in honour of Tiberius, settled in it the poor and the ignorant among the people, the reason being that Tiberias was built on the site of tombs, and others did not wish to settle in it because of its impurity.[8] The capital of Galilee was thus an impure place and the people living in it were accordingly impure.

There are halakhot quoted by some of the above-mentioned scholars

---

4 See above, pp. 4–5.

5 S. Baron, *A Social and Religious History of the Jews*, New York, 1952², I, p. 278.

6 Mark 7:1; cf. also 3:22, where it is stated that at Capernaum there were "the Scribes who came down from Jerusalem."

7 See, for example, G. Dalman, *Orte und Wege Jesu*, Gütersloh, 1924³, p. 120.

8 *AJ*, xviii, 2, 3, §§36–38. And see TB Shabbat 33b–34a.

which testify to an extremely scrupulous observance of the commandments in Judaea but not in Galilee. חמר בתרומה, שביהודה נאמנים על טהרת יין ושמן, כל ימות השנה, ובשעת הגתות והבדים אף על התרומה "A greater stringency applies to terumah [than to holy things], for in Judaea they are trusted in respect of the purity of [holy] wine and oil throughout the year, and only at the season of the wine-presses and the olive-vats also in respect of terumah."[9] From this the Talmud correctly concludes that חומר בתרומה — שביהודה נאמנין, הא בגליל לא " 'a greater stringency applies to terumah — for in Judaea they are trusted' [in respect of it], but not in Galilee."[10] The Mishnah deals with maintaining the purity of the wine for the libations of the altar as well as the purity of the oil for the meal-offerings requiring oil. In maintaining this purity the Judaeans were believed but not the Galilaeans. When the grapes were pressed in the wine-press and the olives in the olive-vat, the Judaeans, but not the Galilaeans, were trusted also in respect of the purity of terumah. Hence these scholars concluded that Galilee was populated by the 'ammei ha-aretz who were not scrupulous in the observance of purity, whereas Judaea was inhabited by ḥaverim and by ne'emanim.

According to a tradition in the Talmud, Rabban Johanan b. Zakkai lived eighteen years during the days of the Second Temple in Galilee in a place called 'Arav.[11] During all this time he was consulted on halakhic problems only twice. Complaining of this, Rabban Johanan b. Zakkai accused Galilee of being devoid of a knowledge of the Torah, and prophesied that because of this, a calamity would befall it. רבי עולא אמר, שמונה עשר שנין עביד הוי יהיב [ריב"ז] בהדא ערב, ולא אתא קומוי, אלא אילין תרין עובדיא, אמר, גליל גליל שנאת התורה, סופך לעשות במסיקין "R. 'Ulla said, Eighteen years he [Rabban Johanan b. Zakkai] lived at 'Arav, and only these two cases came before him, whereupon he exclaimed, Galilee, Galilee, you hate the Torah. Your end will be that you will fall into the hands of conductores."[12] Some connect this with the statement of R. Eliezer b. Hyrcanus, the pupil of Rabban Johanan

---

9 Ḥagigah iii, 4.

10 TJ Ḥagigah iii, 79b; and, similarly, TB Ḥagigah 25a.

11 'Arav, once an important settlement in Lower Galilee, was situated near Bet Netophah valley, about 12 kms. NNE of Sepphoris, today the Arab village of 'Arabe. See A. Schalit, *Namenwörterbuch zu Flavius Josephus*, Leyden, 1968, p. 32, and the bibliography *ad loc.*

12 TJ Shabbat xvi, 15d; מסיקין = מציקים, these are the conductores, the large-scale tenant-farmers, who were given the lands expropriated by the Romans after the Great Revolt against them.

b. Zakkai, who prophesied that הגליל יחרב "Galilee will be destroyed."[13]

There are expressions of disdain of the Galilaeans testifying to their ignorance. So, for example, in the generation of Jabneh, Beruriah, the wife of R. Meir, used the expression גלילי שוטה "foolish Galilaean." רבי יוסי הגלילי הוה קא אזיל באורחא, אשכחה לברוריה, אמר לה, באיזו דרך נלך ללוד, אמרה ליה, גלילי שוטה, לא כך אמרו חכמים, אל תרבה שיחה עם האשה, היה לך לומר, באיזה ללוד "R. Jose the Galilaean was once going along a road, when he met Beruriah. He asked her, Which road does one take to Lydda? She replied, Foolish Galilaean, did not the Sages say, Do not engage in much talk with a woman? You should have said, Which to Lydda?"[14]

Some cite the following tradition as evidence of the non-observance of commandments in Galilee. אמר רבי ישמעאל [בכ״י וינה: רבי שמעון שזורי], מבעלי בתים שבגליל היו בית אבא, מפני מה חרב, מפני שדנו דיני ממונות באחד, ושגידלו בהמה דקה וכו׳ "R. Ishmael [Vienna MS.: 'R. Simeon Shezuri'] said, My father's family was among the householders of Galilee. Why was it destroyed? Because among them civil cases were adjudicated by one judge, and because they reared small cattle..."[15]

These views, as also the proofs advanced in support of them, have been disputed by several scholars of our generation, in particular by Klein[16] and Allon.[17] These scholars point not only to the necessity of a systematic investigation of all the sources referring to Judaea and to Galilee, but also to the weakness of some of the above-mentioned proofs.

The sources which have been cited do not testify to a sparseness of the knowledge of the Torah and of the observance of the commandments in Galilee, nor to its having been inhabited by 'ammei ha-aretz.

The evidences in the New Testament about the Sages who came to

---

[13] Soṭah ix, 15. In the parallel passages the statement is given in the names of Sages of other periods: in the name of R. Judah in TB Sanhedrin 97a; in that of Resh Laḳish in Canticles Rabbah ii, 13; of R. Abbun in Pesiḳta de-Rav Kahana v, "Ha-Ḥodesh ha-Zeh," ed. Mandelbaum, p. 98, and in Pesiḳta Rabbati xv, ed. Friedmann, 75b.

[14] TB 'Eruvin 53b. In this passage there are various traditions, the dates of which cannot in every case be determined and which tell of the Galilaeans' ignorance, in that they were not exact in their Hebrew, and so on.

[15] Tosefta Bava Ḳamma viii, 14.

[16] See S. Klein, *Eretz ha-Galil*, Jerusalem, 1967², *passim*, and, in particular, pp. 33–34; and see also his other writings.

[17] See G. Allon, *Toledot ha-Yehudim be-Eretz Yisra'el bi-Teḳufat ha-Mishnah ve-ha-Talmud*, Tel Aviv, I: 1958³, II: 1961² — see I, pp. 90–91, 318–322; *idem*, *Meḥḳarim be-Toledot Yisra'el*, Tel Aviv, 1957–1958, I, pp. 163–164.

Galilee from Jerusalem do not prove that in Galilee itself there were no Sages.[18] Jerusalem in the days of the Second Temple, as also Jabneh after the destruction of the Temple, were centres of the Torah and the seats of the spiritual-religious leadership. No wonder, then, that the Galilaeans received the Sages who came from these places with honour,[19] but this is no proof that Galilee was occupied by ignorant people and that there were no Sages among its sons.

The account of the founding of Tiberias is to be seen as reflecting an historical situation,[20] and it is precisely this account which demonstrates the care exercised by the majority of the Galilaeans in matters relating to purity, since it shows that Herod Antipas had so much difficulty in populating Tiberias that he had to entice or compel the ignorant and the poor to settle there, for others were unwilling to do so.

The distinction made by the halakhah between Judaea and Galilee as regards those trusted about the purity of wine and of oil was not because of the greater observance of the commandments in Judaea as compared to Galilee, nor because Galilee was populated by 'ammei ha-aretz as against Judaea's talmidei ḥakhamim and ḥaverim. Rather is Safrai's explanation[21] to be accepted that it was due to the much closer association with the Temple which existed in Judaea in consequence of that region's geographical proximity to it. For the one treading out the wine or pressing the oil, the prospect of his taking them himself to the Temple or selling them to someone going on pilgrimage there was sufficiently great to induce him to be scrupulous about the purity of the wine or the oil, even if he himself as an 'am ha-aretz was not generally particular about these things. On the other hand, in Galilee, where such a prospect was slender either because of its distance from the Temple or because the number of pilgrims from Galilee was smaller than in Judaea, the 'ammei ha-aretz were not particular about the purity of their wine or their oil.[22]

---

18 On Galilaean Sages, see below, p. 210.

19 For Sages who came from Jabneh to Galilee, see below, p. 211.

20 Klein, *Eretz ha-Galil*, pp. 100–101, doubts the reliability of this account on the ground that the prohibition against being in an impure place applies only to priests, and thus he follows those who limit to priests only the obligation relating to purity. There is no foundation for Klein's view that Josephus' account is exaggerated and tendentious.

21 See S. Safrai, *Ha-'Aliyah le-Regel bi-Yemei Bayit Sheni*, Tel Aviv, 1965, pp. 44–46, and also p. 157.

22 The explanation in the relevant passages in both TJ and TB Ḥagigah that wine and oil coming from Galilee were not pure, because רצועה של כותים מפסקת "a strip

Proof of this is to be found in the halakhah contained in the continuation of the Mishnah in Ḥagigah which states מן המודיעית ולפנים נאמנין על כלי חרס, מן המודיעית ולחוץ אין נאמנין. כיצד, הקדר שהוא מוכר הקדרות, נכנס לפנים מן המודיעית, הוא הקדר והן הקדרות והן הלוקחים — נאמן, יצא, אינו נאמן "From Modiʻim inwards [the potters] are trusted in respect of earthenware vessels; from Modiʻim outwards they are not trusted. For instance, if the potter who sells the pots enters inwards of Modiʻim, then the same potter in respect of the same pots and of the same buyers is trustworthy, but if he went outwards [of Modiʻim] he is not trusted."[23] It is thus clear that the distinction was not between Judaeans and Galilaeans, but in the relation of the region to the Temple. Thus as regards the very same potter, the very same pots, and the very same buyers, before the potter entered within the limits of Modiʻim he was not trusted in respect of the pots if he declared them to be pure. Once however he entered within the limits of Modiʻim he was trusted. But if he went outside the limits of Modiʻim he was again not trusted.

In general it may be said that these halakhot once more show what has been mentioned several times previously, and that is that the ʻammei ha-aretz were not suspected of deliberately disregarding commandments. It was only that they did not keep, in all their details, the difficult commandments of the tithes and of purity. But when the observance of these commandments was necessary and the ʻammei ha-aretz wanted to keep them, they were not suspected of failing to observe the commandments, but were on the contrary presumed to be trustworthy. What is more, their word was also trusted, as is evident from the continuation of a previously quoted halakhah. ...ובשעת הגתות והבדים [נאמנים] אף על התרומה. עברו הגתות והבדים והביאו לו חבית של יין של תרומה, לא יקבלנה ממנו, אבל מניחה לגת הבאה, ואם אמר לו, הפרשתי לתוכה רביעית קדש נאמן "...And at the season of the wine-presses and the olive-vats [they are trusted] also in respect of the terumah. If [the season of] the wine-presses and the olive-vats has passed and he [the ʻam ha-aretz] brought him [the priest] a jar of terumah wine, the latter may not accept it from him. However he [the ʻam ha-aretz] may leave it for the coming [season of] the wine-press. But if he [the ʻam ha-aretz] said to him [the priest], In it I have set apart a quarter log[a] as a holy thing, he is trusted [in respect of the

---

of [land inhabited by] Cuthaeans separates" Galilee from Judaea, does not reflect the historical background of the halakhah. See Safrai, *op. cit.*, *loc. cit.*

[23] Ḥagigah iii, 5.

a) A liquid measure.

purity of the whole]."[24] Thus, should the owner of the jar, even if an 'am ha-aretz, say to the priest that he had put into it a quarter log of wine which he had kept in purity for the libations of the altar, he is trusted in respect of the purity of the entire jar, including the terumah in it.

The incident related about Rabban Johanan b. Zakkai is also not decisive proof of the sparseness of the observance of the Torah in Galilee, for the following reasons. i) The statement was transmitted by the Amora R. 'Ulla, who belonged to the fifth generation of the Eretz Israel Amoraim, that is, he lived three hundred years and more after the events referred to took place. ii) It is doubtful whether Rabban Johanan b. Zakkai indeed spent eighteen whole years in Galilee, nor are there any additional evidences in support of it. Before the destruction of the Temple, we find him, as a rule, in Jerusalem. iii) The prophecy that Galilee would be full of conductores is surprising, since after the destruction of the Temple conductores increased also in Judaea, where they were apparently more numerous than in Galilee. In view of all this, it is more probable that 'Ulla's statement was an homiletical interpretation of a tradition which he had and which stated that Rabban Johanan b. Zakkai was consulted about two halakhic questions when he was at 'Arav.

Nor do expressions of the type of גלילי שוטה "foolish Galilaean" constitute any proof, since they were used provocatively. We have previously dealt with the acrimonious language usual at that time. Beruriah, who was herself of Galilaean origin [for her father, R. Ḥanina b. Teradion, was the head of the bet din at Sikhni in Galilee during the period of Jabneh], used this expression when speaking to the important and well-known Sage, R. Jose the Galilaean, so that it is no indication of a sparse knowledge of the Torah in Galilee.

Disparaging expressions, similar to those customarily used about the Galilaeans, occur in a tradition of the third century CE about the Daromaeans [the Southerners] and the Babylonians. רבי שמלאי אתא גבי רבי יונתן, אמר ליה אלפן אגדה, אמר ליה מסורת בידי מאבותי, שלא ללמד אגדה לא לבבלי ולא לדרומי, שהן גסי רוח ומעוטי תורה "R. Simlai came to R. Jonathan and said to him, Teach us aggadah. He replied, I have a tradition from my forefathers not to teach aggadah either to a Babylonian or to a Daromaean, for they are haughty and have a scant knowledge of the

---

24 Ḥagigah iii, 4.

Torah."[25] It is impossible to maintain that at the time this statement was made there was no knowledge of the Torah in Darom [the South], seeing that important Sages lived there, such as R. Joshua b. Levi at Lydda. The same applies to Babylonia, whose emergence as a centre of the Torah had at that time already begun. Just as the expression בבלאי טפשאי "stupid Babylonians,"[26] frequently used in the Talmud, does not indicate a lack of a knowledge of the Torah in Babylonia, so the term "foolish Galilaean" does not denote the absence of a knowledge of the Torah in Galilee. These terms are to be seen as no more than trite phrases in which, at the expense of the Galilaeans, the Judaeans expressed their local patriotism.

Nor do the reasons given by R. Simeon Shezuri [this is the probable version] for the destruction of Galilee refer to the disregard there of important commandments. The prohibition against rearing small cattle, to which we have previously referred, was a regulation introduced by the Sages of Jabneh with the object of safeguarding the economy and the settlements. Neither was judgment in civil cases by a single judge fundamentally opposed to the halakhah. It was only that the Sages of Jabneh viewed it unfavourably. This is one of the sources which shows that Galilee took part in the Bar Kokheva war. Thus the inhabitants of Galilee made common cause with the Judaeans in the revolt, which was supported by the Sages. It is precisely this source which indicates that there was a close connection between Galilee and Judaea, rather than that they were remote from each other in character.[27]

A systematic examination of the sources reveals that in Galilee there were observed in the days of the Second Temple and in the generation of Jabneh those areas which the ‘ammei ha-aretz are said to have disregarded. These sources do not of course indicate that all the Galilaeans were particular in such matters, but they clearly demonstrate that Galilee cannot be described as having been inhabited entirely by ‘ammei ha-aretz.

---

25 TJ Pesaḥim v, 32a.

26 TB Pesaḥim 34b, and parallel passages; TB Betzah 16a, and parallel passages; TB Ketubbot 75a; TB Nedarim 49b, and parallel passages.

27 On manifestations of revolt in Galilee and the destruction that followed in its wake at the time of the Bar Kokheva revolt, cf. also TJ Ta‘anit iv, 69a; Lamentations Rabbah i, 45; Pe’ah vii, 1; TJ Pe’ah vii, 20a; Tosefta Shevi‘it vii, 18; TB Bava Batra 75b; TB Yevamot 96b. See also a letter from Bar Kokheva in *DJD*, II, 160.

Tithes

In his autobiography Josephus tells that his fellow-priests, Joazar and Judas, who were sent with him to Galilee at the beginning of the Great Revolt against the Romans, amassed a large sum of money from the tithes and wanted to return home.[28]

Elsewhere in his autobiography Josephus praises his own conduct, in that, despite his rights as a priest, he did not take from those who brought them the tithes which were due to him.[29]

In several parallel passages in talmudic literature a letter is cited which was sent by the heads of the Sanhedrin to both the Galilaeans and the Judaeans, instructing them לאפוקי מעשריא "to bring out the tithes":[30] מעשה ברבן גמליאל וזקנים שהיו יושבין על גב מעלות בהר הבית ויוחנן סופר הלה לפניהם, אמרו לו כתוב לאחנא בני גלילא עילאה ולבני גלילא תחתאה, שלמכון יסגא, מהודענא לכון די מטא זמן ביעורא לאפוקי מעשריא ממעוטני זיתייא. ולאחנא בני דרומא עילאה ובני דרומא תחתא שלמכון יסגא, מהודענא לכון אנחנא דומטא זמן ביעורא לאפוקי מעשריא מעומרי שבלייא וכו׳ "Once when Rabban Gamaliel and the elders were sitting on the steps on the Temple Mount and Johanan the scribe was standing in front of them, they said to him, Write to our brethren in Upper Galilee and to those in Lower Galilee, May your peace be great. We wish to inform you that the time for the removal has come to bring out the tithes from the olive heaps. And to our brethren in Upper Daroma [South] and to those in Lower Daroma, May your peace be great. We wish to inform you that the time for the removal has come to bring out the tithes from the sheaves of corn..."[31] Here we have a living picture of the separation of the tithes in both Galilee and Judaea.

For the period of Jabneh, too, there is evidence of the acceptance in Galilee of the obligation to separate the tithes. Thus the people of Meron asked R. Akiva a practical halakhic question relating to tithes. אמר רבי יוסי, מעשה שנכנסה גרן אחת של פול במירון, ובאו ושאלו את רבי עקיבא, והתיר להן את השוק, אמר להן, נתמעט וחזר השוק ליושנה "R. Jose said, It happened that a crop of beans [from outside Eretz Israel] came into Meron, and they approached R. Akiva and asked [him about it]. He allowed them [to use] the market [without separating tithes], saying to

[28] *Li e*, xii, §§62–63.

[29] *Ibid.*, xv, §80.

[30] The precise significance of the Hebrew is not quite clear.

[31] Tosefta Sanhedrin ii, 6; and cf. TJ Sanhedrin i, 18d; TB Sanhedrin 11b; Midrash Tannaim to Deut. 26:13, p. 176.

them, When [the crop of beans] has diminished, the market returns to its original state [and is liable to tithes].''[32]

The Sabbatical Year

An incident that occurred in the generation of Jabneh demonstrates the adherence in Galilee to the observance of the Sabbatical Year. אמר רבי יוסי, מעשה שזרעו כרם גדור בציפורי בצלים, למוצאי שביעית זרעוהו שעורים, והיו פועלים יורדין ומנכשין לתוכו, ומביאין ירק לתוך קופותיהם, ובאו ושאלו את רבי יוחנן בן נורי, ואמר, אם עשו כיוצא בהן מותרין, ואם לאו אסורין "R. Jose said, Once a fenced vineyard in Sepphoris was sown with onions, and on the termination of the Sabbatical Year it was sown with barley. Workers went down, weeded it, and bringing vegetables in their baskets, came and asked R. Johanan b. Nuri, who said, If a new crop of the same kind has ripened, they [the vegetables] are permitted; if not, they are prohibited.''[33] Accordingly, in the generation of Jabneh there were observed in Sepphoris not only the Sabbatical Year, but also the restrictions which were added by the Sages and which prohibited the taking of vegetables on the termination of the Sabbatical Year until a new crop of the same kind had ripened.[34]

Purity and Impurity

It has previously been stated that Josephus' account of the founding of Tiberias, if accepted as an historical fact, shows that the majority of the Galilaeans were scrupulous in the observance of purity and refused to settle in Tiberias because of the impurity of the tombs there.

Several sources in talmudic literature testify to the Galilaeans' scrupulous observance in various areas of purity.

A circumstance which is apparently to be ascribed to the period of Jabneh demonstrates that the people of Sepphoris were particular that their vegetables should not be rendered liable to impurity. אמר רבי יוסי, בראשונה היו הבכרות של קשואין ושל דלועין שבצפורי טמאות, מפני שמקנחין אותו בספוג בשעה שעוקרין אותן, קבלו עליהן אנשי ציפורי, שלא יהו עושין כן "R. Jose said, At first the early-ripening cucumbers and pumpkins in Sepphoris were impure because, when picking them, they cleaned them with

[32] Tosefta Demai iv, 13.

[33] Tosefta Shevi'it iv, 13.

[34] Cf. Shevi'it vi, 4; and see S. Safrai, "Sabbatical Year Commandments under the Conditions Prevailing after the Destruction of the Second Temple" (Hebrew), *Tarbiz*, 35 (1966), pp. 311–312.

a sponge, [until] the people of Sepphoris took it upon themselves not to do so."[35] The reference here, it should be emphasized, is to the desire of the people of Sepphoris to observe the purity of secular food.

The people of Rum Bet 'Anat[36] in Galilee asked R. Ḥanina b. Teradion of Sikhni, of the generation of Jabneh, for a practical halakhic decision on a problem relating to a ritual bath. שוב מעשה ברום בית ענת שקוות יותר מאלפים כור, ובאו ושאלו את רבי חנניה בן תרדיון, ופסל שאני אומר נכנסו גויים וזלפוה בלילה וחזרו ומלאו אותה בקילון "It also happened at Rum Bet 'Anat that more than two thousand kor [a measure of capacity] of water were gathered [in the ritual bath], and they came and asked R. Ḥanina b. Teradion, who disqualified [it] on the ground that non-Jews had entered and emptied it at night and refilled it with a water hoist [thus making it drawn water which renders a ritual bath unfit]."[37]

The Torah and its Study

Nor can Galilee in the days of the Second Temple and in the generation of Jabneh be described as devoid of the Torah and as the exclusive abode of the 'ammei ha-aretz la-Torah.

There were important Sages in the days of the Second Temple and in the generation of Jabneh who were Galilaeans. Some of them were the heads of batei midrash in Galilee, some taught the Torah in Galilee, others again went to the centres of the Sages and of the leadership — to Jerusalem in the period of the Second Temple and to Jabneh after the destruction of the Temple.

There were, for example, R. Jose the Galilaean and R. Ḥalafta of Sepphoris, R. Ḥanina b. Teradion, the head of the bet din at Sikhni,[38] and Abba Jose Ḥoliḳofri of Ṭiv'on,[39] and many others.

There were also Sages from Galilee who went to teach the Torah in Judaea. Thus we find R. Jose the Galilaean in Darom [the South]. In the generation of Jabneh, Galilaeans were to be found in the main batei midrash in Judaea, such as in those at Jabneh, Lydda, and Benei Beraḳ. But not only the Sages were in contact with Judaea and its batei mid-

---

35 Tosefta Makhshirin iii, 5.

36 It is usually identified as a place in Galilee, apparently south of Bet Netophah valley (but S. Klein, *Sefer ha-Yishuv*, Jerusalem, 1939, I, s.v., locates it SW of Damascus; see also *idem*, *IEJ*, I, 3 [1933], pp. 4–7).

37 Tosefta Miḳva'ot vi, 3.

38 Sikhni = Sikhnin, north of Bet Netophah valley. And see TB Sanhedrin 32b: אחר רבי חנניא בן תרדיון לסיכני "Follow R. Hananiah (Ḥanina) b. Teradion to Sikhni."

39 Makhshirin i, 3.

rash. To the centre at Jabneh came the people of Sepphoris to inquire about an halakhah relating to kil'ayim [the prohibition against mingling heterogeneous seeds or plants]. בשקי של ציפרי היו מרכיבין קרוסטמל על גבי עוגס, מצאן תלמיד אחד, אמר להן אסורין אתם, הלכו וקצצום, ובאו ושאלו ביבנה, אמרו מי פגע בכם, אינו אלא מתלמידי בית שמאי "In an irrigated field of Sepphoris they grafted Crustumenian on native pear-trees. A scholar, coming across this, told them, You are not allowed [to do it, whereupon] they went and cut them down. They came and asked at Jabneh, where they said, Who has done you a disservice? It can only be one of Bet Shammai's pupils."[40] It should be added that when R. Akiva's son died, they came to lament him from both Galilee and Judaea.[41]

There were Sages from Judaea who came to Galilee to teach the Torah there. A frequent statement in the Talmud is הריני כבן עזאי בשוקי טבריא "I am like Ben ʿAzzai in the market places of Tiberias."[42] A well-known Tanna of the generation of Jabneh, Ben ʿAzzai was, it is clear from this statement, active in Tiberias. It is similarly told of Elisha b. Avuyah, who was likewise of the generation of Jabneh, that היה יושב ושונה בבקעת גיניסר "he sat and taught the Torah in the valley of Gennesar,"[43] obviously before he became an apostate. Of R. Eliezer b. Hyrcanus, one of the principal Sages of Jabneh, it is told in a Baraita ששבת בגליל העליון, ושאלוהו שלשים הלכות בהלכות סוכה "that he passed the Sabbath in Upper Galilee and they asked him thirty halakhot in the laws of the sukkah."[a] [44] These and similar sources testify to the spread of the Torah in Galilee and to its inhabitants' thirst for knowledge both in theoretical studies and in the solution of practical halakhic problems.

40 Tosefta Kil'ayim i, 4. The statement is further evidence of the observance of commandments in Galilee.

41 Semaḥot viii, 13, ed. Higger, pp. 159–161; and see TB Moʿed Ḳaṭan 21b.

42 TB ʿEruvin 29a; TB Soṭah 45a; TB Ḳiddushin 20a; TB ʿArakhin 30b. It is against this background that we are to understand the statement of R. Johanan, the av bet din of Tiberias in the third century CE: רבי יוחנן אזל לחד אתר, אמר אנא בן עזאי דהכא "R. Johanan came to a place and said, I am the Ben ʿAzzai of this place" (TJ Bikkurim ii, 65a).

43 TJ Ḥagigah ii, 77b.

44 TB Sukkah 28a. In a parallel passage reference is made to תלמיד אחד מתלמידי גליל העליון "one of the pupils of Upper Galilee" who mentioned to R. Eleazar an halakhah he had heard (Tosefta Kelim, Bava Metziʿa ii, 1; TB Shabbat 52b); cf. also the incident in Tosefta Meʿilah i, 5.

a) The booth for the festival of Tabernacles (see Lev. 23:42).

Supervisory Visits of the Nasi and the Sanhedrin, and the Sanhedrin's Local Sessions, in Galilee in the Generation of Jabneh

One of the ways in which the leadership institutions exercised their authority in the period of Jabneh, and in subsequent periods too, was by paying visits throughout the country. We find the Nasi, together with the most important Sages, visiting settlements in the country, discussing with their inhabitants halakhic questions, and supervising the arrangements in them. Such visits, which included settlements alike in Judaea and in Galilee, reveal that there was no national-spiritual difference between the Galilaeans and the Judaeans.

For example, אמר רבי יהודה מעשה בשגביון ראש בית הכנסת של אכזיב שלקח כרם רבעי מן הגוי בסוריא, ונתן לו דמיו, ובא ושאל את רבן גמליאל שהיה עובר ממקום למקום וכו׳ "R. Judah said, Segavyon, the head of the synagogue of Achzib, once bought a vineyard in its fourth year from a non-Jew in Syria, paid him for it, and then came and asked Rabban Gamaliel, who was going from one place to another..."[45] In this instance the Nasi arrived at Achzib in western Galilee in the course of a visit, during which he "went from one place to another."

Another example: נגר שיש בראשו גלסטרא, רבי אלעזר אוסר ורבי יוסי מתיר. אמר רבי אלעזר, מעשה בכנסת שבטבריא שהיו נוהגין בו התר, עד שבא רבן גמליאל והזקנים ואסרו להן. רבי יוסי אומר, אסור נהגו בה, בא רבן גמליאל והזקנים והתירו להן "If a bolt has a knob at one end, R. Eleazar forbids it [to be moved on the Sabbath], but R. Jose permits it. R. Eleazar said, In a synagogue at Tiberias the common practice was to treat it as permitted, until Rabban Gamaliel and the elders came and forbade it to them. R. Jose said, They treated it as forbidden, but Rabban Gamaliel and the elders came and permitted it to them."[46]

We similarly find the leadership visiting Acco [Acre][47] and Kefar ʿUtnai.[48] There was also the occasion when the Nasi and the heads of the Sanhedrin were in the עיירות של כותים "cities of the Cuthaeans,"[49] apparently when on their way from Judaea to Galilee.[50] As mentioned

---

45 Tosefta Terumot ii, 13.

46 ʿEruvin x, 10. For the presence of the Nasi Rabban Gamaliel of Jabneh at Tiberias, see also Tosefta Shabbat xiii, 2.

47 ʿAvodah Zarah iii, 4; Tosefta Moʿed Ḳaṭan ii, 15.

48 Giṭṭin i, 5. Kefar ʿUtnai was on the southern border of Lower Galilee.

49 Tosefta Demai v, 24; and see below, pp. 231–232.

50 According to I. Halevy, *Dorot ha-Ri'shonim*, Pressburg-Frankfort-on-Main, 1901–1918, III, p. 348, the journeys of the Nasi and of the accompanying Sages from place to place in Galilee do not refer to supervisory visits but to their escape from

previously, similar visits also took place in Judaea, and we find the Nasi, together with the Sages accompanying him, at Ashkelon, Jericho, and Benei Beraḳ, and likewise at other settlements. To some extent such visits were also paid to the diaspora.

A distinction may be drawn between two types of visits by the Nasi and the Sages, the one, supervisory visits, when the Nasi and the Sages accompanying him went from one settlement to another, the other, when the Nasi and the Sages with him remained for some time in a settlement where they held a local session of the Sanhedrin.

The following source, which refers to Usha in Galilee, is apparently to be explained in the context of such a local session. וכבר שלחה מלכות שני סרדיטיאות ואמרה להם לכו ועשו עצמכם יהודים, וראו תורתם מה טיבה, הלכו להם אצל רבן גמליאל לאושא "The government of Rome once sent two commissioners, saying to them, Go and make yourselves Jews, and see what is the nature of their Torah. They went to Rabban Gamaliel at Usha."[51] [52]

Additional Restrictions in Galilee as Compared to Judaea

A number of sources bear witness to a difference in the customs of Galilee and of Judaea in halakhic and other areas. Some of these sources refer to incidents which took place during the leadership's supervisory visits in Galilee.

For example, יושבין על ספסל של גוים בשבת, בראשונה היו אומרין אין יושבין על ספסל של גוים בשבת. מעשה ברבן גמליאל שהיה יושב על ספסל של גוים בשבת בעכו, אמרו לו, אין נוהגין כן להיות יושבין על ספסל של גוים בשבת, לא רצה לומר להם מותרין אתם, אלא עמד והלך לו "One may sit on the bench of non-Jews on the Sabbath. At first they said, One may not sit on the bench of non-Jews on the Sabbath. It happened that Rabban Gamaliel sat on the bench of non-Jews on the Sabbath at Acco [Acre], and they said to him, It is not customary here to sit on the bench of non-Jews on the Sabbath. Since he did not wish to tell them that it was permitted for them [to do so], he got up and walked away."[53] In this instance

---

the persecutions that took place in the days of Domitian. This is however not substantiated by the sources. Apparently Rabban Gamaliel became Nasi after the death of Domitian in 96 CE, when there are certainly no evidences of persecutions.

51 Sifrei Deuteronomy cccxliv, ed. Finkelstein, p. 401.

52 The view that the Sanhedrin moved to Usha as early as in the days of Rabban Gamaliel of Jabneh is improbable. See Allon, *Toledot ha-Yehudim*, I, pp. 292–293.

53 Tosefta Moʿed Ḳaṭan ii, 14–15. According to the Vienna and London MSS., the lenient view was introduced in the period of Jabneh itself by R. Akiva.

the Galilaeans observed the ancient, stringent halakhah, which laid down that it was prohibited to sit on the Sabbath on a bench of non-Jews since it was used for business purposes, as it might have created the impression that the Jew was engaged in trading on the Sabbath. While this may, it is true, indicate that innovations in the halakhah had not yet reached Galilee, it nevertheless attests to the strict observance of the halakhah there.

In incidents unconnected with the leadership's visits we also find an excessive strictness in the observance of the halakhah by the Galilaeans, as, for example, ביהודה היו עושין מלאכה בערבי פסחים עד חצות, ובגליל לא היו עושין כל עקר "In Judaea they used to work on the eve of Passover until midday, whereas in Galilee they did not work [then] at all."[54]

In various spheres the customs of the Galilaeans were more beautiful than those of the Judaeans.

After the incident of "the bench of non-Jews," other incidents are quoted which occurred in various places in Galilee and in which the sons of the Nasi were involved. מעשה ביהודה והילל בניו של רבן גמליאל, שנכנסו לרחוץ בכבול, אמרו להן, אין נוהגין כאן להיות רוחצין שני אחין כאחד, ולא רצו לומר להם מותרים אתם, אלא נכנסו ורחצו זה אחר זה. שוב מעשה ביהודה והילל בניו של רבן גמליאל, שהיו יוצאין בקורדקייסין בשבת בבירי, אמרו להם אין נוהגין כן להיות יוצאין בקורדקייסין בשבת, ולא רצו לומר להם מותרין אתם, אלא שלחום ביד עבדיהם "It once happened that Judah and Hillel, the sons of Rabban Gamaliel, entered [a bath-house] at Cabul to have a bath. They [the people] said to them, It is not customary here for two brothers to have a bath together. Not wishing to tell them that it was permitted for them [to do so], they went in and bathed one after the other. It also happened that Judah and Hillel, the sons of Rabban Gamaliel, went out at Biri in slippers on the Sabbath. They [the people] said to them, It is not customary here to go out in slippers on the Sabbath. Not wishing to tell them that it was permitted for them [to do so], they sent them [the slippers] on with their servants."[55]

---

[54] Pesaḥim iv, 5.

[55] Tosefta Moʻed Ḳaṭan ii, 15–16. Cabul — the present-day Arab settlement of the same name, about 15 kms. NE of Acco [Acre]. Biri — apparently Biriyya of today, one km. north of Safed. See also Tosefta Shabbat vii, 17. Tosefta Ahilot xvi, 12 tells of how Judah and Hillel, the sons of Rabban Gamaliel, decided a point of law for the Galilaeans. מעשה ביהודה והלל אחין בניו של רבן גמליאל, שהיו מהלכין בתחום עוני. מצאו אדם אחד שנפחת קברו בתוך שדהו, אמרו לו מלקט עצם עצם והכל טהור "Once when Judah and his brother Hillel, the sons of Rabban Gamaliel, were walking in the region of ʻOni, they came across a man in whose field a grave had

Among the Galilaeans there was also a more beautiful custom relating to the ketubah than among the Judaeans. את תהא יתבא בביתי ומתזנא מנכסי, כל ימי מגד אלמנותיך בביתי – חיב, שהוא תנאי בית דין. כך היו אנשי ירושלים כותבין, אנשי הגליל היו כותבין כאנשי ירושלים. אנשי יהודה היו כותבין, עד שירצו היורשין לתן ליך כתבתיך, לפיכך אם רצו היורשין, נותנין לה כתבתה ופוטרין אותה "[Similarly if he did not give his wife the written undertaking], 'You shall dwell in my house and shall be maintained therein out of my estate throughout the duration of your widowhood,' he is nevertheless liable, since [also this clause] is a condition laid down by the bet din. So did the men of Jerusalem write. The men of Galilee wrote in the same manner as the men of Jerusalem. The men of Judaea however used to write, 'Until the heirs may consent to pay your ketubah.' Consequently, the heirs may, if they so wish, pay her ketubah and dismiss her."[56] Thus the Galilaeans, adopting the custom of Jerusalem, secured in the ketubah the right of the widow to maintenance and to living in the home of her late husband as long as she wished, whereas the Judaeans enabled the heirs to free themselves of this obligation by giving the widow the money stipulated in the ketubah. Accordingly the Jerusalem Talmud declared אנשי הגליל חסו על כבודן ולא חסו על ממונן, אנשי יהודה חסו על ממונן ולא חסו על כבודן "The men of Galilee cared for their honour and did not care for their money; the men of Judaea cared for their money and did not care for their honour."[57]

In other areas, too, there were more beautiful customs in Galilee than in Judaea.[58]

To sum up. The evidences as a whole show that the Jewish character of Galilee was not less than that of Judaea. Nor is there any justification for associating the concepts of the ʻam ha-aretz le-mitzvot and the ʻam ha-aretz la-Torah with Galilee to the exclusion of Judaea.

The distance from the centres — from Jerusalem and the Temple during the existence of the latter, and from Jabneh after its destruc-

---

fallen in, whereupon they said to him, Gather bone by bone, and everything is pure." It is difficult to identify תחום עוני "the region of 'Oni": the reference may be to the region of Acco [so Rash — R. Samson of Sens]. See also Lieberman, *Tosefet Ri'shonim*, III, p. 147. Here we have further evidence of a scrupulous observance of purity in Galilee.

56 Ketubbot iv, 12.

57 TJ Ketubbot iv, 29b.

58 In the area of ethics — TB Shabbat 153a; Semaḥot iii, 6. In comforting mourners — TJ Berakhot ii, 5b; etc.; cf. also TB Ketubbot 16b. In matters relating to marriage — Tosefta Ketubbot i, 4; TJ Ketubbot i, 25a; TB Ketubbot 12a.

tion — may of course have had an effect in certain spheres, including those in which the ʿam ha-aretz was not considered to be strictly observant. But this did not stem from an inferior national-spiritual approach on the part of the Galilaeans but rather from their slender geographical link with the centres.

Due also to their distance from the centres, the Galilaeans were less up to date in their knowledge of the development of the halakhah than the Judaeans. Since frequently this development was precisely towards a more lenient practice, the stricter view was in such instances followed in Galilee rather than in Judaea.

All the evidences relating to the observance of the commandments and the study of the Torah in Galilee refer, it must be emphasized, alike to the small settlements, to Tiberias and Sepphoris, the main cities of Galilee, and to both Lower and Upper Galilee. This is not to suggest that all the Galilaeans were talmidei ḥakhamim who were scrupulous about the observance of a lenient as well as of a stricter halakhah, or that in the settlements everyone was equally observant of the commandments and equally learned in the Torah. Nonetheless the evidences reveal a line steadfastly followed in the observance of the halakhah and the commandments and in the study of the Torah in Galilee as a whole.

To refute the view of scholars, and especially that of Büchler, we have cited in particular evidences indicating that the Galilaeans were scrupulous in observing things about which the ʿam ha-aretz was said to be not particular. In the course however of our treatment of the subject, examples have also been given of the general observance of commandments unconnected with the question of the ʿam ha-aretz, such as the Sabbath, kil'ayim, and so on, and there are further evidences which we have not quoted.

Mention should also be made of the participation of Galilee in the Bar Kokheva revolt. Here too things are symptomatic. Although the main centre of the revolt and its principal objectives were in Judaea, the Galilaeans did not absolve themselves from the obligation imposed on them by the war. Evidences may be found in the available sources of revolt, and naturally of ensuing destruction, also in Galilee.

When after the Bar Kokheva revolt the centre moved to Galilee, we do not hear of difficulties about "the judaization of Galilee," as contended by Büchler. The contrary is the case. The leadership, which was reconstituted at Usha in Galilee, included not only Sages who arrived from Judaea, but also Galilaean Sages, such as R. Judah b. Ilaʿi who was himself from Usha, as well as Sages who came from Babylonia, such as R. Nathan who occupied the position of av bet din. Within a short time the leadership institutions succeeded in establishing in Galilee a stable Jewish-national life, despite the grave crisis which followed the Bar Kokheva revolt. All this would presumably have been impossible had Galilee been until that time גליל הגויים "Galilee of the nations."

After the Bar Kokheva revolt Galilee became the centre of the Torah and of the nation, with Judaea occupying a secondary position. This similarly gave rise to the view that the world of the Torah and of the commandments was henceforth to be found only in Galilee and no longer in Judaea. But even as in the days of the Second Temple and in the period of Jabneh it was wrong to speak of Galilee as devoid of the Torah and of commandments, so was there no foundation for such a view about Judaea after the Bar Kokheva revolt. For there continued to be centres of the Torah in Judaea, as, for example, at Lydda; in the days of R. Judah ha-Nasi the association of "the Holy Congregation in Jerusalem" was founded; there were groups of Sages in Judaea, such as זקני הדרום "the elders of Darom [the South]."

In general, the study of the Torah and the observance of the commandments were to be found alike in Judaea and in Galilee during the entire period of the Mishnah and of the Talmud. When the centre was in Judaea, Galilee drew its sustenance from it, and vice versa. And alongside the Sages and those observant of the commandments in Judaea and Galilee, ʻammei ha-aretz were during all this period to be found both in Judaea and in Galilee.

CHAPTER SEVEN

THE 'AMMEI HA-ARETZ, THE CHRISTIANS, AND THE SAMARITANS

1. The 'Ammei ha-Aretz and the Emergence of Christianity

We have several times referred to the view, which was current among scholars, that Christianity emerged from, and was nurtured by, the world of the 'ammei ha-aretz. Jesus came to fight their battle. Striving in a reformative tendency on behalf of the 'ammei ha-aretz, he opposed the conflict that prevailed between the Pharisaic Sages and the ḥaverim on the one hand and the ordinary people on the other, a conflict which was the outcome of the former's deliberate social shunning of the latter.[1] This view we have rejected completely. And yet there are several points of contact between the 'ammei ha-aretz and emergent Christianity, some imaginary, others actual. With these we propose to deal.

The New Testament expresses sharp opposition to the commandments which were scrupulously observed by the ḥaverim and the Pharisees, and for the neglect of which the 'ammei ha-aretz were condemned — the commandments of the tithes and of purity.

A condemnation of the Pharisees and the Scribes for being particular about the observance of the commandments of the tithes and of purity, while neglecting the principles of the Torah, occurs in the following passage. "Woe to you, Scribes and Pharisees, hypocrites! for you tithe mint and dill and cummin, and have neglected the weightier matters of the law, justice and mercy and faith; these you ought to have done, without neglecting the others. You blind guides, straining out a gnat and swallowing a camel! Woe to you, Scribes and Pharisees, hypocrites! for you cleanse the outside of the cup and of the plate, but inside they are full of extortion and rapacity. You blind Pharisee! first cleanse the inside of the cup and of the plate, that the outside also may be clean. Woe to you, Scribes and Pharisees, hypocrites! for you are like white-washed tombs, which outwardly appear beautiful, but within they are full of dead men's bones and all uncleanness. So you also outwardly

[1] See especially pp. 2 and 200, above.

appear righteous to men, but within you are full of hypocrisy and iniquity."[2]

Elsewhere it is related that the Pharisees and the Scribes who came from Jerusalem gathered together to Jesus when he was in Gennesaret and rebuked him because his disciples ate bread without washing their hands, as is customary among all Jews. Jesus' reply was similar to that quoted above, in that he remonstrated with the Pharisees who preferred the observance of the tradition rather than of the central commandments of the Torah. "Now when the Pharisees gathered together to him, with some of the Scribes, who had come from Jerusalem, they saw that some of his disciples ate with hands defiled, that is, unwashed. [For the Pharisees, and all the Jews, do not eat unless they wash their hands, observing the tradition of the elders; and when they come from the market place, they do not eat unless they purify themselves; and there are many other traditions which they observe, the washing of cups and pots and vessels of bronze.] And the Pharisees and the Scribes asked him, 'Why do your disciples not live according to the tradition of the elders, but eat with hands defiled?' And he said to them, 'Well did Isaiah prophesy of you hypocrites, as it is written,

> "This people honours me with their lips, but their heart is far from me; in vain do they worship me, teaching as doctrines the precepts of men."

You leave the commandment of God, and hold fast the tradition of men.' And he said to them, 'You have a fine way of rejecting the commandment of God, in order to keep your tradition! For Moses said, "Honour your father and your mother"; and, "He who speaks evil of father or mother, let him surely die"; but you say, "If a man tells his father or his mother, What you would have gained from me is Corban" [that is, given to God] — then you no longer permit him to do anything for his father or mother, thus making void the word of God through your tradition which you hand on. And many such things you do.' And he called the people to him again, and said to them, 'Hear me, all of you, and understand: there is nothing outside a man which by going into him can defile him; but the things which come out of a man are what defile him.' And when he had entered the house, and left the people, his disciples asked him about the parable. And he said to them, 'Then are you also without understanding? Do you not see that whatever goes into a man from outside cannot defile him, since it enters,

[2] Matthew 23:23–28.

not his heart but his stomach, and so passes on?' [Thus he declared all foods clean.] And he said, 'What comes out of a man is what defiles a man. For from within, out of the heart of man, come evil thoughts, fornication, theft, murder, adultery, coveting, wickedness, deceit, licentiousness, envy, slander, pride, foolishness. All these evil things come from within, and they defile a man'."[3]

In general it may be said that there was some exaggeration on the part of the evangelists when stating that on these issues Jesus set himself against all the Sages and against "all the Jews." The passage cited above deals mainly with the washing of the hands for the eating of secular food, and we have previously mentioned that this custom prevailed among individuals, among ḥaverim in associations, and among Pharisees and those close to them, but under no circumstances was it a custom of all the Jews.[4] Thus, for example, the washing of the hands was included among the various stages in the process of admission to the association of ḥaverim,[5] and hence it cannot be said that it was customary among all Israel. Accordingly, there is no occasion here to set the Christians and the 'ammei ha-aretz on one side of the barricade and "all the Jews" on the other.

Nor was opposition to stringencies in the observance of purity limited to Christians, for it was also to be found, though less vehemently of course, among "those peaceable and faithful in Israel."

In some of their customs the ḥasidim resembled "the associations in Jerusalem," in that they, too, emphasized action in all that pertained to benevolence, social help, and so on. But in their customs no trace is to be found of their trying to set themselves apart from, and exalt themselves above, the community, or of their seeking to adopt strin-

---

[3] Mark 7:1–23. Cf. Matthew 15:1–6; and similarly Romans 14:14; etc. The New Testament statements are not in every case contrary to the halakhah in general. Thus, for example, the statement in Mark about a vow which vitiates the commandment of honouring one's father and mother is compatible with the view of R. Eliezer [in the spirit of Bet Shammai ?] but opposed to that of the Sages [Bet Hillel ?], and finally even מודים חכמים לר׳ אליעזר בדבר שבינו לבין אביו ואמו שפותחין לו בכבוד אביו ואמו "the Sages agreed with R. Eliezer that in a matter concerning himself and his father and his mother, the honour due to them is suggested as an opening [for the absolution of a vow]" (Nedarim ix, 1; cf. v, 6; xi, 4; and see J. N. Epstein, *Mevo'ot le-Sifrut ha-Tanna'im*, Jerusalem, 1957, pp. 377–378).

[4] G. Allon, *Meḥḳarim be-Toledot Yisra'el*, Tel Aviv, 1957–1958, I, pp. 164–169, who holds that the eating of secular food in a state of purity was not an additional restriction, nevertheless emphasizes that it was not customary among the people.

[5] See, in particular, pp. 121–124, above.

gencies in matters relating to purity, but on the contrary, they tended rather to a lack of strictness in such things.

The following tradition attests to the scant familiarity of ḥasidim with the subject of purity. ודלא שמש חכימייא קטלא חייב. מעשה בכהן אחד חסיד ברמת בני ענת, והלך רבי יהושע לדבר עמו והיו עוסקים בהלכות חסידים, וכיון שהגיעה עונתה של סעודה אמר לאשתו, הביאי טיפה של שמן לתוך הגריסין. הלכה ונטלה את הפך מתוך הכירה. א״ל רבי, וכי הכירה טהורה היא, אמר לו וכי יש כירה טמאה ותנור טמא. א״ל, והרי הוא אומר ״תנור וכירים יותץ טמאים הם,״ הא שיש תנור טמא וכירים טמאים ודשמש ולא קיים חייב קטלי קטילין. אמר, רבי כך הייתי נוהג כל ימי, אמר לו אם כך היית נוהג כל ימיך לא היית אוכל קדשי שמים כתקונן ''And whoever has not ministered to the Sages deserves death. There was once at Ramat Benei ʻAnat a priest who was a ḥasid. R. Joshua went to speak to him, and they dealt with the halakhot of ḥasidim. When it came to mealtime, he said to his wife, 'Put a drop of oil in the beans,' and she went and took the jar from the stove. R. [Joshua] said to him, 'Is the stove pure?' He answered him, 'Are there, then, an impure stove and an impure oven?' He said to him, 'Does it not say, ''Whether oven or stove, it shall be broken in pieces; they are impure.'' So there are an impure oven and an impure stove, and he who has ministered [to the Sages] and does not observe, deserves death.' He answered, 'Rabbi, so I have been accustomed to do all my days.' 'If this is what you have been accustomed to do all your days,' he replied, 'you have not eaten the holy things of heaven properly'.''[6]

This incident also occurs in Version A of Avot de-R. Nathan in a parallel passage which contains several supplementary and explanatory details. מעשה באדם אחד מבית רמה שהיה נוהג בעצמו מידת חסידות. שלח אליו תלמיד אחד רבן יוחנן בן זכאי לבודקו, הלך ומצאו שנטל שמן ונתנו על גבי כירים ונטלו מעל הכירים ונתנו לתוך מקפה של גריסין. א״ל מה אתה עושה. א״ל כהן גדול אני ותרומה בטהרה אני אוכל. א״ל כירים זה טמא או טהור... א״ל אם כן היית נוהג לא אכלת תרומה טהורה מימיך ''There was once at Bet Ramah a man who conducted himself after the manner of ḥasidim. Rabban Johanan b. Zakkai sent a pupil to him to examine him and he went and found that he took oil and put it on the stove and took it from the stove and put it in the dish of pounded beans. He said to him, 'What are you doing?' He answered, 'I am a high priest and eat terumah in purity.' He said to him, 'Is this stove pure or impure?'... 'If this is what you

6 Avot de-R. Nathan, Version B, xxvii, ed. Schechter, p. 56.

have been accustomed to do,' he replied, 'then you have never eaten pure terumah'."[7]

The expression שלח... לבודקו "he sent... to examine him" occurs in the sources in the sense of examining a person to find out whether he was not to be presumed to be a heretic and a dissentient from the ways of the ḥalakhah.[8] If in the above incident we divest the historical kernel of its literary wrappings, we find that there was a need to examine the ḥasidim on the nature of their ḥalakhot,[9] so that what we have here is presumably not a simple incident, the purpose of which is to tell of a priest who was a ḥasid and who did not know a biblical verse relating to a central subject in the sphere of purity and impurity. Rather does this incident show that the ḥasidim, even priestly ones, were not scrupulous in the observance of purity and that this did not meet with the approval of the Sages.[10]

Nor did the Sages themselves in every instance represent the extreme line in the observance of purity. It was none other than Sages who protested against an excessive scrupulousness in this sphere, so that we find them criticizing the inordinate attention paid by priests to purity and impurity. Thus, for example, מעשה בשני כהנים, שהיו שוין ורצין ועולין בכבש, דחף אחד מהן את חבירו לתוך ארבע אמות, נטל את הסכין תקע לו בליבו. בא רבי צדוק ועמד על פתח האולם בהר הבית, אמר להן שמעוני אחינו בית ישראל, הרי הוא אומר "כי ימצא חלל באדמה וגו׳ ויצאו זקניך ושופטיך," בואו ונמדוד על מי ראוי להביא את העגלה על ההיכל או לעזרות, געו כל העם בביכיה, ואחר כך בא אביו של תינוק ואמר להן, עדיין בני מפרפר ולא נטמאת הסכין, מלמד שטומאת הסכין היה קשה להן יותר משפיכות דמים, וכן הוא אומר "וגם דם נקי שפך מנשה הרבה מאוד," מיכן אמרו, בעון שפיכות דמים שכינה נעלת ובית המקדש מיטמא "It

---

[7] Avot de-R. Nathan, Version A, xii, ed. Schechter, p. 56; and cf. Midrash ha-Gadol Lev., MS., to Lev. 11:35, ed. Rabinowitz, pp. 241–242.

[8] So, for example, in the following case: ...ומעשה באחד שהיה כותב את הברכות, אמרו עליו לפני רבי ישמעאל, הלך רבי ישמעאל לבודקו וכו׳ "...And there was the case of one who wrote out the blessings. They told R. Ishmael about him. R. Ishmael went to examine him..." (Tosefta Shabbat xiii, 4; and cf. TJ Shabbat xvi, 15c; TB Shabbat 115b; Massekhet Soferim xv, 4, ed. Higger, p. 276). In this case, it is clear from the context, R. Ishmael went to examine the person to find out whether he was not to be suspected of heresy.

[9] R. Joshua, it should be noted, on various occasions carried out missions on behalf of Rabban Johanan b. Zakkai with the purpose of inquiring into the actions of the public (see, for example, Derekh Eretz Rabbah vi, Pirḳei Ben 'Azzai iv, 1, ed. Higger, pp. 193–200).

[10] And see S. Safrai, "Teaching of Pietists in Mishnaic Literature," *JJS*, 16 (1965), pp. 15–33.

once happened that two priests were equal as they ran and mounted the ramp leading to the altar. When they were within four cubits [of the altar], the one pushed the other, took the knife, and thrust it into his heart. R. Zadok came and standing at the entrance of the Vestibule on the Temple Mount, said to them, 'Hearken to me, our brethren, the house of Israel. It is said, "If in the land... any one is found slain... then your elders and your judges shall come forth, [and they shall measure the distance to the cities which are round about... and the elders of the city which is nearest to the slain man shall take a heifer...]" Come, let us measure on whose behalf a heifer should be brought, whether on behalf of the Sanctuary or of the Temple courts.' [Thereupon] all the people burst out crying. Later the father of the young man came and said to them, 'My son is still in his death-throes, and the knife has not been made impure.' Thus the impurity of the knife was harder for them than the shedding of blood. And so it says, 'Moreover Manasseh shed very much innocent blood.' Accordingly they declared, For the sin of the shedding of blood, the divine presence has withdrawn, and the Temple has been rendered impure."[11]

Some Sages, it should be added, were themselves not scrupulous about purity and impurity. So, for example, there is the previously cited account concerning Simeon b. Nethanel the priest who was presumably R. Simeon b. Nethanel the priest, one of Rabban Johanan b. Zakkai's pupils:[12] מעשה ברבן גמליאל הזקן שהשיא את בתו לשמעון בן נתנאל הכהן ופסק עמו על מנת שלא תעשה טהרות על גביו "It was told of Rabban Gamaliel the Elder that he married his daughter to Simeon b. Nethanel the priest and stipulated with him that it was on condition that she was not to prepare food for him in ritual purity."[13] It should also be mentioned that talmidei ḥakhamim were obliged to pass all the stages of admission to the association of the ḥaverim, and that most of these stages were connected with purity and impurity.[14] We have seen that Resh Laḳish said עם הארץ אני אצל הטהרות "I am an 'am ha-aretz as regards ritual purities,"[15] for already in his days there had set in a general laxity in the observance of purity and impurity.[16]

---

[11] Tosefta Yoma i, 12.

[12] Avot ii, 8; Avot de-R. Nathan, Version B, xiii, ed. Schechter, p. 31.

[13] Tosefta 'Avodah Zarah iii, 10.

[14] See above, pp. 120–131.

[15] TJ Demai ii, 23a; and see S. Lieberman, *Tosefet Ri'shonim*, II, pp. 190–191; and *idem*, "Further Notes on the Leiden MS. of the Jerushalmi" (Hebrew), *Tarbiz*, 20 (1949), p. 110.

[16] See above, pp. 66, 89.

From what has been said above it follows that in the Judaism of both the Second Temple and the tannaitic periods there was a wide spectrum of views and patterns of living in everything pertaining to the observance of purity. There were even, as we have seen, Sages and ḥasidim who protested against an excessive scrupulousness as regards purity and some of whom were themselves not particular about its observance. It is thus far-fetched to assume that there was a conceptual and social identity between the early Christians and the ʻammei ha-aretz, merely because Christians expressed opposition to the way in which purity was observed by Pharisees, and the ʻammei ha-aretz were condemned for not being particular about the observance of purity. We have moreover sought to show that the ʻammei ha-aretz were not opposed to the observance of purity but simply disregarded it, while in a considerable number of cases and halakhot we have found that the ʻammei ha-aretz were considered trustworthy in one or another matter relating to purity.[17]

There is no substance in the entire approach which sets the ʻammei ha-aretz le-mitzvot and the Pharisees on opposite sides of the barricade. We have in various instances come across the existence of close relations between the Pharisees and the ʻammei ha-aretz, and have even seen how the ʻammei ha-aretz made common cause with the Pharisees, so that we certainly cannot talk of an estrangement between them.[18] Accordingly, any assumption about close relations and even an identity between the Jewish-Christian community and the ʻammei ha-aretz is based from the very outset on shaky foundations.

Among the definitions of an ʻam ha-aretz are those contained in the following statement. רבי יהושע אומר, כל שאינו מניח תפילין. בן עזאי אומר, כל שאין לו ציצית בבגדו "R. Joshua said, Anyone who does not put on tefillin. Ben ʻAzzai said, Anyone who does not have tzitzit on his garment."[19] From the New Testament as well as from the Church Fathers we learn of opposition to the observance of these commandments. "Then said Jesus to the crowds and to his disciples, 'The Scribes and the Pharisees sit on Moses' seat; so practise and observe whatever they tell you, but not what they do; for they preach, but do not practise. They bind heavy burdens, hard to bear, and lay them on men's shoulders; but they themselves will not move them with their finger. They do all their

---

17 See above, pp. 90–92.

18 See above, pp. 157–159, 161–169.

19 TB Berakhot 47b; and cf. TB Soṭah 22a.

deeds to be seen by men; for they make their phylacteries broad and their fringes long'."[20] On the basis of these verses, too, there would appear to be a connection between the Christians and the ʻammei ha-aretz.

Even in the generations following the destruction of the Second Temple, the commandments of the tefillin and the tzitzit had not yet spread to such an extent that the early Christians could be contrasted with the people as a whole, so that it could be maintained that if the ʻammei ha-aretz also did not practise the commandments of the tzitzit and the tefillin, it necessarily followed that they belonged to the Christian sect.

In the period after the destruction of the Second Temple, the commandments of the tzitzit and the tefillin crystallized, as did the others which the ʻam ha-aretz was said to disregard,[21] in a group of Baraitot in tractate Berakhot. Sages dealt with the halakhot pertaining to the tzitzit and tefillin,[22] and incidents are recorded about the obligation to observe these commandments in practice.[23] But, as happened also in other cases, these commandments did not spread immediately among the entire nation. The following statement by R. Simeon b. Eleazar shows that not everyone was particular about observing the commandment of the tefillin. כל מצוה שלא מסרו ישראל עצמן עליה למיתה בשעת גזרת המלכות כגון תפילין, עדיין היא מרופה בידם "Every commandment for which the Jewish people did not risk their lives during persecutions, such as [the commandment of] the tefillin, is still laxly observed by them."[24] In later generations, too, there were continuing complaints about the non-observance of the commandment of the tefillin by many of the nation,[25] and presumably a similar situation obtained with regard to the observance of the commandment of the tzitzit.

The emphasis on the great importance of these commandments and the demand that they be scrupulously observed indicate that not everyone was scrupulous in observing them. There is, for example, the statement of R. Meir. וכן היה רבי מאיר אומר, אין לך אדם מישראל שאין מצות מקיפות אותו, תפילין בראשו ותפילין בזרועו ומזוזה בפיתחו וארבעה ציציות מקיפות

20 Matthew 23:1–5; and cf. Justin, *Dialogue with Tryphon*, 46.

21 The mezuzah, the recital of the shemaʻ, the study of the Torah, ministering to talmidei ḥakhamim.

22 For example, in TB ʻEruvin 96a; TB Moʻed Ḳaṭan 21a; TJ Pe'ah i, 15b; TB Menaḥot 32b, 42b.

23 TJ Berakhot ii, 4c; TB Sanhedrin 68a, 92a; TB ʻEruvin 96a; etc.

24 TB Shabbat 130a.

25 For example, TJ Berakhot ii, 4c.

אותו, ועליהן אמר דוד "שבע ביום היללתיך על משפטי צדקך וכו'" "And thus said R. Meir, There is no Jew who is not surrounded by commandments, [by] tefillin on his head, and tefillin on his arm, and a mezuzah at his door, and four tzitziyyot encompassing him. And of them David said, 'Seven times a day I praise thee for thy righteous ordinances...'."[26] It is interesting to note that whereas R. Meir generally defined the ‘am ha-aretz as one who did not eat his secular food in a state of purity,[27] here he called for the observance of those commandments to which the Sages of the generation of Jabneh and of his own generation resorted when defining the ‘am ha-aretz.[28] A discussion in tractate Shabbat deals with the sins responsible for a person's death. Among the views expressed is also that of R. Meir who maintained that it was בעון מזוזה "for the sin of neglecting the commandment of the mezuzah," while R. Judah held that it was בעון ציצית "for the sin of neglecting the commandment of the tzitzit."[29] Commendation by R. Meir of anyone who properly observed the commandment of the tzitzit occurs in the following source. רבי מאיר אומר, וראיתם אותם לא נאמר כאן, אלא "וראיתם אותו," מגיד הכתוב שכל המקיים מצות ציצית מעלים עליו כאלו הקביל פני שכינה, שהתכלת דומה לים וים דומה לרקיע, והרקיע דומה לכסא הכבוד, כענין שנאמר "וממעל לרקיע אשר על ראשם כמראה אבן ספיר דמות כסא" "R. Meir said, It is not stated here 'that you may look upon them' but 'that you may look upon it.' Thus the verse declares that whoever observes the commandment of the tzitzit is accounted as though he had greeted the divine presence, for the blue [of the cord in the tzitzit] resembles the colour of the sea, and the colour of the sea resembles that of the sky, and the colour of the sky resembles that of the throne of glory, as it is said, 'And above the firmament over their heads there was the likeness of a throne, in appearance like sapphire'."[30]

---

26 Tosefta Berakhot vii, 25.

27 TB Berakhot 47b; and parallel passages. Cf. also תני בשם רבי מאיר, כל מי שהוא קבוע בארץ ישראל, ואוכל חוליו בטהרה... מובטח לו שהוא מחיי העולם הבא "R. Meir taught, Whoever is permanently settled in Eretz Israel, and eats his secular food in ritual purity... is assured of a life in the world to come" (TJ Shabbat i, 3c; and parallel passages).

28 Cf. also ת"ר איזהו עם הארץ, כל שאינו קורא קריאת שמע שחרית וערבית בברכותיה, דברי רבי מאיר "Our Rabbis taught, Who is an ‘am ha-aretz? Whoever does not recite the shema‘ morning and evening with its accompanying benedictions; such; is the statement of R. Meir" (TB Soṭah 22a).

29 TB Shabbat 32b.

30 Sifrei Numbers cxv, ed. Horovitz, p. 126; and, similarly, TB Menaḥot 43b.

The stress laid on the importance of observing these commandments and the threat of punishment for their non-observance, which occur in many other sources, are evidence that the observance of these commandments had to be encouraged, since not everyone kept them. In not one of such sources is the non-observance of these commandments connected specifically with Christians. Nor is it to be assumed that the ʿammei ha-aretz did not keep these commandments because they were opposed to them, as were the Christians. Rather was it because of a lack of sufficient scrupulousness in observing them. The definition, given by some Sages, of an ʿam ha-aretz as one who neglected to observe these commandments was in fact a means whereby the Sages sought to enlarge the observance of these very commandments and make them the possession of all Jews. It is in this context that we have to view the following statement. וכן היה רבי שמעון בן אלעזר אומר משום רבי מאיר, מעשה באשה אחת שנישאת לחבר והיתה קומעת על ידיו תפילין, נישאת למוכס וקשרה על ידיו קשרים "And thus R. Simeon b. Eleazar said in the name of R. Meir, It happened that a certain woman, who was married to a ḥaver, fastened the straps of the tefillin for him, and when afterwards she was married to a publican, she knotted the customs seals for him."[31] [32]

In general it may be said that it is doubtful whether the statements in the New Testament express opposition in principle to the commandments of the tzitzit and the tefillin, or opposition to an excessive emphasis on achieving righteousness through the observance of ceremonial commandments.

A further reason for accusing the ʿammei ha-aretz of not observing these commandments is that they all either directly or indirectly refer to the study of the Torah and emphasize its importance. The above-mentioned Baraitot apparently relate to the concept of the ʿam ha-aretz la-Torah which, we believe, came into being after the destruction of the Second Temple. Hence the definition of the ʿam ha-aretz as one who did not observe the commandments of the tzitzit, the tefillin, reciting the shemaʿ, or the mezuzah was not because the ʿam ha-aretz disregarded specifically these commandments and kept others, but because they are in one way or another connected with the subject of the study of the Torah.[33]

---

[31] Tosefta Demai ii, 17; and, similarly, TB Bekhorot 30b.

[32] See A. Büchler, *Der galiläische ʿAm-haʾAreṣ des zweiten Jahrhunderts*, Vienna, 1906, pp. 22–25, and the notes *ad loc.*

[33] See above, p. 97.

It is accordingly clear that there is no basis for connecting the Jewish-Christian sects with the ʻam ha-aretz la-Torah. General verses, such as, "Have any of the authorities or the Pharisees believed in him? But this crowd, who do not know the law, are accursed,"[34] constitute no foundation for associating the ʻam ha-aretz la-Torah with Christianity, especially since there is no evidence that the Pharisees held themselves aloof from the masses of the people, nor is the concept of the ʻam ha-aretz to be identified specifically with the masses.

In the initial period after the destruction of the Second Temple, the fate of the Christians was, from a certain point of view, similar to that of the ʻammei ha-aretz. It was only later that the fate of the one differed from that of the other.

At the end of the Second Temple period the Christians were regarded as one of the many Jewish sects that existed in those days and as part of the Jewish people. There are evidences of a positive attitude by Pharisaic Sages to the Christians in the period of the Second Temple.[35] After its destruction however there developed a tendency among the Sages at the head of the people to reject the Christian sects from the totality of the nation. This tendency found expression in practice in the institution of the benediction relating to heretics,[36] in various actions,[37] and in statements like that of R. Ṭarfon: שאפילו אדם רודף אחריו להורגו, ונחש רץ להכישו, נכנס לבית עבודה זרה, ואין נכנס לבתיהן של אלו [בתי מינים] "Even if someone runs after him to kill him, or a snake pursues him to bite him, he should enter a heathen temple but not the homes of these [heretics]."[38]

Among the various reasons that led to this tendency[39] the principal one was apparently the Sages' desire to unite the people and to exclude from the community all sections which in spirit and in action were not

---

[34] John 7:48–49.

[35] Thus, for example, Rabban Gamaliel the Elder sided with the apostles when some of them were brought before the Sanhedrin for sentence (Acts 5:33 ff.; and cf. 23:9); the Pharisees were embittered at the killing of James, the brother of Jesus, by a Sadducaean court (*AJ*, xx, 9, 1, §§197–203).

[36] TJ Berakhot iv, 8a; TB Berakhot 28b.

[37] Tosefta Ḥullin ii, 22–24, and parallel passages; TB Sanhedrin 74a, and parallel passages.

[38] TB Shabbat 116a; and see the entire discussion there.

[39] For the subject as a whole, see G. Allon, *Toledot ha-Yehudim be-Eretz Yisra'el bi-Teḳufat ha-Mishnah ve-ha-Talmud*, Tel Aviv, I: 1958[3], II: 1961[2] — see I, pp. 178–192.

in full accord with the leadership of the nation and with the nation itself.

A similar fate befell the 'ammei ha-aretz. And this is the reason to which we have attributed the hatred of the 'ammei ha-aretz that had its inception in the generation of Jabneh. Because the 'ammei ha-aretz did not join the Sages in their efforts, following the destruction of the Second Temple, to make the Torah and its study the way of life of the people, the Sages were bitterly opposed to them, even to the extent of introducing prohibitions, to which we have previously referred, against establishing family ties or close relations with them, or reposing any confidence in them in a matter calling for some measure of responsibility.[40]

In the end however there was a difference in the fate of the Christians and the 'ammei ha-aretz. Whereas the Jewish-Christian sects were totally rejected by Judaism, after several generations the opposition to the 'ammei ha-aretz subsided. This divergent development was the outcome of the basic difference between them. The Christian sects developed their own outlook on the world, a hope of salvation inconsistent with the beliefs and views of the Sages, a different interpretation of the Pentateuch and of the Prophets, and a faith in Jesus as the redeemer and teacher vested with a special status. By contrast, the 'ammei ha-aretz le-mitzvot or la-Torah were not a sect with its own principles but rather a social stream that had neither institutions nor frameworks, and that was not scrupulous in the observance of all the commandments and in the study of the Torah in accordance with the outlook of the Sages and the teachers of the Torah. The 'ammei ha-aretz did not deny Judaism, neither were they opposed in principle to the halakhah as a whole, so that when suitable conditions arose they were received back into the body of Judaism with open arms.[41]

2. The 'Ammei ha-Aretz and the Samaritans

There is a similarity between the 'ammei ha-aretz and the Samaritans in a number of areas, and in several halakhic sources they are mentioned side by side.

Tithes

Tractate Kutim lays down unequivocally ופירותיהן טבל כפירות הגויים

[40] See, in particular, pp. 172–176, above.

[41] See above, pp. 92–96, 161–169, 189–195.

"And their produce is ṭevel,[a)] like the produce of gentiles."[42] This is either a theoretical halakhah or one that reflects a later situation, for in the tannaitic period there are evidences of an uncertainty about the produce of gentiles, even as there was about the produce of an 'am ha-aretz.[43]

The different views of Tannaim on the subject of the tithing of the Samaritans' produce are to be found in the following source. מעשרין משל ישראל על של גוים, ומשל גוים על של ישראל, ומשל ישראל על של כותי, ומשל כותי על של ישראל, ומן הכל על הכל, דברי רבי מאיר. רבי יהודה, רבי יוסי ורבי שמעון אומרים, מעשרין של ישראל על של ישראל, ושל גוים על של גוים, ושל כותי על של כותי, אבל לא מישראל על של גוים ועל של כותי, ולא מהן על של ישראל... רבי אליעזר אומר, בשל כותים כשם שעשו פירות ישראל דמאי אחר רובן אין מעשרין מזה על זה, כך מעשרין פירות כותי דמאי, ואין מעשרין מזה על זה "One may give tithes from produce bought from a Jew for other produce bought from a gentile, and from produce bought from a gentile for other produce bought from a Jew, and from produce bought from a Jew for other produce bought from a Cuthaean,[b)] and from produce bought from a Cuthaean for other produce bought from a Jew, and from produce bought from any of these for other produce bought from any one of them. This is the view of R. Meir. R. Judah, R. Jose, and R. Simeon said, One may separate tithes from the produce bought from a Jew for other produce bought from a Jew, and from produce bought from a gentile for other produce bought from a gentile, and from produce bought from a Cuthaean for other produce bought from a Cuthaean, but not from produce bought from a Jew for other produce bought from a gentile and for other produce bought from a Cuthaean, nor from produce bought from them for other produce bought from a Jew... R. Eliezer said, As regards the produce of Cuthaeans, even as the produce of a Jew was made demai following the majority of them and tithes may not be separated from produce bought from one Jew for other produce bought from another Jew, so tithes are separated from the produce of a Cuthaean as demai and tithes may not be separated

---

42 Kutim i, 7, ed. Higger, p. 62.

43 It is not our intention to deal with the question whether a Cuthaean was like a Jew or a gentile, except in so far as this is relevant to the concrete problems connected with the concept of the 'am ha-aretz.

a) Produce at the stage at which tithes and terumot should be, but have not yet been, separated.

b) Term used by the Sages when referring to the Samaritans.

from produce bought from one Cuthaean for other produce bought from another Cuthaean."[44]

R. Meir, consistent with his line of thought, held that a gentile could not own property in Eretz Israel so fully as to release his produce from the obligation to tithe it. Hence the produce of Cuthaeans was ṭevel. Tithes were therefore separated from produce bought from a Jew for other produce bought from a Cuthaean, and from produce bought from a Cuthaean for other produce bought from a Cuthaean. In the opinion however of R. Judah, R. Jose, and R. Simeon, the Cuthaeans were not obliged to separate tithes at all, and hence tithes were not separated from produce bought from a Jew for other produce bought from Cuthaeans. R. Eliezer held that Cuthaeans were obliged to separate tithes, and may even have tithed some of their produce, which accordingly had the status of demai, like the produce of an 'am ha-aretz. Thus tithes were separated from the produce of a Cuthaean as demai, but were not separated from produce bought from one Cuthaean for other produce bought from another Cuthaean.

Another dispute relating to the Cuthaeans and tithes is to be found in the following incident. מעשה שנכנסו רבותינו לעיירות של כותים שעל יד הדרך, הביאו לפניהם ירק, קפץ רבי עקיבא ועישרן ודאי. אמר לו רבן גמליאל, היאך מלאך ליבך לעבור על דברי חביריך, או מי נתן לך רשות לעשר, אמר לו, וכי הלכה קבעתי בישראל, אמר לו, ירק שלי עישרתי, אמר לו, תדע שקבעתה הלכה בישראל שעישרתה ירק שלך, וכשבא רבן גמליאל ביניהם, עשה תבואה וקטנית שלהן דמאי, ושאר כל פירותיהם וודאי, וכשחזר רבן שמעון בן גמליאל ביניהן, ראה שנתקלקלו ועשר כל פירותיהן וודאי "Once when our Rabbis entered the cities of the Cuthaeans beside the way, vegetables were brought before them, whereupon R. Akiva jumped up and tithed them as being definitely untithed. Rabban Gamaliel said to him, 'How is it that you have presumed to disregard the words of your colleagues? And who gave you permission

[44] Tosefta Demai v, 21–22; and, similarly, TJ Demai v, 24d–25a; TB Menaḥot 66b (where, in the printed versions, R. Judah holds like R. Meir and not like R. Jose and R. Simeon; and see Diḳduḳei Soferim); see Lieberman, *Tosefta ki-Peshuṭah*, Seder Zera'im, p. 260. Cf. Demai v, 9, where R. Eleazar holds the view quoted in the Tosefta in the name of R. Eliezer, but in the Mishnayot in the Babylonian Talmud, in the Munich MS., in Maimonides' Mishnah commentary, etc., the reading is R. Eliezer; and see J. N. Epstein, *Mavo le-Nusaḥ ha-Mishnah*, Jerusalem, 1964[2], II, p. 1177. For some trustworthiness on the part of the Cuthaeans with regard to tithes, see also Tosefta Pe'ah iv, 1; and cf. Kutim i, 7, ed. Higger, p. 62. See also TJ Demai vi, 25a, 25d; Tosefta Demai vii, 11. In Samaritan literature the commandment relating to the tithes occurs as an obligatory commandment: see, for example, the Samaritan Book of Joshua xxxviii in Kirchheim, *Karmei Shomeron*, pp. 75–76.

to separate the tithes?' 'Have I, then,' he answered, 'laid down an halakhah in Israel? I have tithed my own vegetables.' 'Know,' replied Rabban Gamaliel, 'that by tithing your vegetables, you have laid down an halakhah in Israel.' And when Rabban Gamaliel came among them, he declared their grain and pulse demai, and the rest of their produce to be definitely untithed. And when Rabban Simeon b. Gamaliel returned among them and saw that they had degenerated, he declared all their produce to be definitely untithed.''[45] [46]

Underlying this incident is the question of the trustworthiness of the Cuthaeans concerning tithes. Not one of the Sages doubted but that they were obliged, even as Jews were, to separate tithes. Thus again we find a similarity between the Cuthaeans and the 'ammei ha-aretz in the view expressed here that the Cuthaeans' produce was not to be tithed as though it were definitely untithed, and that at least their grain and pulse were demai. The account of the incident concludes with the later enactment which laid down that all the produce of the Cuthaeans was to be regarded as definitely untithed. This apparently reflects a change which took place in the Cuthaeans' trustworthiness as regards separating tithes, since it is emphasized that the halakhah was modified because the Cuthaeans ''had degenerated.'' The historical background to this change may lie in the relations engendered between the Cuthaeans and the Jews in connection with the Bar Kokheva revolt.[47]

It is interesting to note that the Babylonian Talmud preserves a tradition which states that the Cuthaeans, in contrast to the 'ammei ha-aretz, were trusted about the tithes. **דתניא איזהו עם הארץ... חכמים אומרים כל שאינו מעשר פירותיו כראוי. והני כותאי עשורי מעשרי כדחזי, דבמאי דכתיב באורייתא מזהר זהירי, דאמר מר, כל מצוה שהחזיקו בה כותים הרבה מדקדקין בה יותר מישראל** ''It has been taught, Who is an 'am ha-aretz?... The Sages said, Anyone who does not tithe his produce properly. Now these Cuthaeans do tithe their produce in the proper way, since they are very scrupulous about any injunction written in the Torah. For a Master

---

[45] Tosefta Demai v, 24. The printed versions have לעיירות של כותים שעל יד הירדן ''to the cities of the Cuthaeans beside the Jordan,'' and כשחזר רבן גמליאל לביניהן ''When Rabban Gamaliel returned among them.'' Cf. Tosefta Demai i, 11.

[46] This incident reflects the tension which existed between Rabban Gamaliel and R. Akiva. Cf. Tosefta Berakhot iv, 15; Tosefta Betzah ii, 12; Tosefta Demai ii, 24; Ro'sh ha-Shanah i, 6; TJ Ro'sh ha-Shanah i, 57b; TB Ro'sh ha-Shanah 22a; Sifra, Ḳedoshim, iv, 9, ed. Weiss, 89b.

[47] See Allon, *Toledot ha-Yehudim*, II, pp. 24–26, as also the sources and bibliography *ad loc.*

has said, Whenever the Cuthaeans have adopted a commandment, they are much more particular about it than the Jews.''[48] This tradition probably reflects the situation which existed before the Bar Kokheva revolt, when the Sages ascribed to the Cuthaeans some trustworthiness in the separation of the tithes, although the view that they were completely trusted about tithes is undoubtedly somewhat exaggerated.

A similarity between Cuthaeans and ʻammei ha-aretz in respect of tithes is to be found in the following source. המוליך חטים לטוחן כותי או לטוחן עם הארץ, בחזקתן למעשרות ולשביעית, לטוחן נכרי, דמאי. המפקיד פרותיו אצל הכותי או אצל עם הארץ, בחזקתן למעשרות ולשביעית, אצל הנכרי כפרותיו, רבי שמעון אומר, דמאי "If a man took his wheat to a miller who was a Cuthaean or to a miller who was an ʻam ha-aretz, [the wheat when ground continues] in its former condition in respect of tithes and the laws of the Sabbatical Year produce. [But if he took it] to a miller who was a gentile, [the wheat when ground becomes] demai. If a man left his produce in the keeping of a Cuthaean or of an ʻam ha-aretz, [it continues when returned] in its former condition in respect of tithes and of the laws of the Sabbatical Year produce. [But if he left it] with a gentile, [it becomes] like the produce of the gentile. R. Simeon said, [It becomes] demai.''[49] The Samaritan miller and bailee were not suspected of changing the wheat or the produce. Thus, even as in the case of the ʻam ha-aretz, a desire deliberately to impair anything relating to tithes was not imputed to the Cuthaeans.

The Sabbatical Year

Several sources indicate that the Cuthaeans were suspected of not observing the laws relating to the Sabbatical Year. For example, שביעית, ישראל משמטין וגוי פטור, ישראל וגוים רבים על הכותים "As regards the Sabbatical Year, Jews are obliged to let the ground lie fallow, but a gentile is exempt [from doing so]. Jews and gentiles are in the majority as compared to the Cuthaeans.''[50] This statement occurs in the context of a discussion that mentions several species in the region of Caesarea which בשביעית היתר, בשאר שני שבוע דמאי "are permitted during the

---

[48] TB Berakhot 47b; and see Diḳduḳei Soferim. For the first part of this passage, see below, p. 237.

[49] Demai iii, 4; and cf. Tosefta Demai iv, 24; TJ Demai iii, 23c; and see Lieberman, *Tosefta ki-Peshuṭah*, Seder Zeraʻim, pp. 240–241. For a reference to both ʻammei ha-aretz and Samaritans in connection with the tithes, see also TJ Demai ii, 22c.

[50] TJ Demai ii, 22c.

Sabbatical Year and are demai during the remainder of the seven year period." This halakhah laid down that most of the produce in the region under discussion was exempt from the laws of the Sabbatical Year or came from a Jew who observed the laws of the Sabbatical Year properly, and that the Cuthaeans constituted a minority that was suspected of not observing the laws relating to the Sabbatical Year. Hence, in the view of the Sages, the Cuthaeans were obliged, even as they were in the case of the tithes, to observe the Sabbatical Year, but were suspected of not doing so.[51]

The halakhah, which we have quoted in connection with the tithes and which likens the trustworthiness of the Cuthaeans to that of the ʻammei ha-aretz in that they would not change wheat and produce deposited with them, is also stated in connection with the Sabbatical Year.

Connected with the subject of the Sabbatical Year is the further problem whether a part of the Land of the Cuthaeans was regarded as heathen country which had never been conquered. It was in relation to this that the following tradition was presumably expressed. עיירות של כותים שנהגו בהן היתר מימי יהושע בן נון והן מותרות "The cities of the Cuthaeans in which, since the days of Joshua the son of Nun, they have been accustomed to permit [Sabbatical Year produce], and which are permitted [during the Sabbatical Year]."[52] According to this view, some Samaritan settlements were exempted from the laws of the Sabbatical Year.[53] However, the relevant passage in the Jerusalem Talmud could be interpreted as referring, not to Cuthaeans, but to Jews who lived in "the cities of the Cuthaeans" and who were not subject to the obligations of the Sabbatical Year. The reason for this may have been that the authorities did not exempt the inhabitants of the Land of the Cuthaeans from taxes during the Sabbatical Year. In this passage, it should be added, stress is laid on the obligation of ḥallah [the priest's share of the dough], and hence it is probable that the entire passage deals with Jews and not with Samaritans.[54]

---

[51] Cf. also Tosefta Shevi'it vi, 20; Lieberman, *Tosefta ki-Peshuṭah*, Seder Zera'im, p. 565. See however Tosefta Shevi'it i, 4, from which it follows that the Samaritans, like the gentiles, were exempted from the requirements of the Sabbatical Year.

[52] TJ Shevi'it vi, 36c.

[53] On this subject, see Allon, *Meḥkarim*, II, pp. 9-10.

[54] The question of tax exemption during the Sabbatical Year has been dealt with by many. See, for example, S. Safrai, "Sabbatical Year Commandments under the Conditions Prevailing after the Destruction of the Second Temple" (Hebrew), *Tarbiz*,

Purity and Impurity

Like the 'ammei ha-aretz, the Samaritans were not scrupulous about the laws of purity and impurity.

Several halakhot deal with the Samaritans' untrustworthiness in observing the laws of niddah [menstruation]. So, for example, בנות כותים נדות מעריסתן, והכותים מטמאים משכב תחתון כעליון מפני שהן בועלי נדות, והן יושבות על כל דם ודם וכו' "The daughters of the Cuthaeans are regarded as menstruant-impure from their cradle, and the Cuthaeans impart impurity to the couch underneath even as [they do] to the upper cover, since they cohabit with menstruants, because [their wives] continue [impure for seven days] on account of the discharge of any blood..."[55]

In the sphere of purity and impurity a Cuthaean was also generally not suspected of deliberately leading astray anyone who wished to observe purity, and in this respect, too, the attitude to them resembled that adopted towards the 'ammei ha-aretz. For example, כל הכתמין הבאים מרקם, טהורין. רבי יהודה מטמא, מפני שהם גרים וטועין. הבאין מבין הגוים, טהורין. מבין ישראל, ומבין הכותים, רבי מאיר מטמא, וחכמים מטהרין, מפני שלא נחשדו על כתמיהן "All bloodstains from Reḳem are pure. R. Judah declared them impure, because the people who live there are proselytes but misguided. Those [bloodstains] that come from the gentiles are pure, those that come from Jews or from Cuthaeans, R. Meir declared impure, whereas the Sages declared them pure, since they are not under suspicion as regards their stains."[56] It should be noted that we have here another halakhah, one of many, in which R. Meir adopted the stricter view in the sphere of purity.

A separate problem, which does not however concern our present work, is the question of the purity of the Land of the Cuthaeans. When we dealt with the subject of the 'ammei ha-aretz in Judaea and Galilee, we rejected the statements of the Talmuds in explaining the Mishnah in Ḥagigah. חמר בתרומה, שביהודה נאמנים על טהרת יין ושמן כל ימות השנה "Greater stringency applies to terumah [than to holy things], for in Judaea they are trusted with regard to the purity of [holy] wine and

35 (1966), pp. 304–328; 36 (1967), pp. 26–46, and the bibliography *ad loc.* See also Allon, *op. cit.*, *loc. cit.* [quoted in previous note].

55 Niddah iv, 1; and cf. iv, 2; Tosefta Niddah v, 1. See Niddah vii, 4–5; Epstein, *Mevo'ot le-Sifrut ha-Tanna'im*, p. 507.

56 Niddah vii, 3. On the problem of the trustworthiness of the Samaritans with regard to purity, see also Oholot xvii, 3; and cf. Niddah vii, 5; Tosefta Shevi'it iii, 13; see also Tosefta Niddah vi, 16, from which the Cuthaeans' trustworthiness in respect of the impurity of graves is derived.

oil throughout the year.'' This is interpreted in the Talmuds as follows: ביהודה אין ובגליל לא... מפני שרצועה של כותים מפסקת ביניהן ''In Judaea, but not in Galilee... because a strip of [land inhabited by] Cuthaeans separates them.''[57] We associate ourselves with the view of those scholars who maintain that the halakhah considered the Land of the Cuthaeans pure, like other parts of Eretz Israel.[58]

The Study of the Torah and the Attitude to the Sages

Not many evidences are extant that would permit a comparison to be drawn or an analogy to be made between the Samaritans and the ‘ammei ha-aretz la-Torah. Nor presumably did the Sages seek to include the Samaritans in the process of making the life of every Jew in Israel centre round the Torah. At the same time there are some evidences of a condemnation of anyone who did not engage in the study of the Torah and did not hearken to the words of the Sages, and of his status being compared to that of a Cuthaean as though he belonged to the circle of the ‘ammei ha-aretz.

Such a parallel between the ‘am ha-aretz le-mitzvot and the Cuthaean is to be found in the following source. אתמר, קרא ושנה ולא שימש תלמידי חכמים, רבי אלעזר אומר, הרי זה עם הארץ, רבי שמואל בר נחמני אמר, הרי זה בור, רבי ינאי אומר, הרי זה כותי, רב אחא בר יעקב אומר הרי זה מגוש ''It has been taught, If one has learnt Scripture and Mishnah but has not ministered to talmidei ḥakhamim, R. Eleazar said, He is an ‘am ha-aretz, R. Samuel b. Naḥmani said, He is an uncultured person [בור], R. Yannai said, He is a Cuthaean, R. Aḥa b. Jacob said, He is a magician, [repeating words he does not understand].''[59]

---

57 Ḥagigah iii, 4; TB Ḥagigah 25a; TJ Ḥagigah iii, 79b–c; and see above, pp. 201–202, 204–205.

58 On the purity of the Land of the Cuthaeans, see Tosefta Miḳva'ot vi, 1; TJ ‘Avodah Zarah v, 44a–b; Tosefta Ahilot xviii, 6; Kutim i, ed. Higger, p. 61. See also S. Safrai, *Ha-‘Aliyah le-Regel bi-Yemei Bayit Sheni*, Tel Aviv, 1965, pp. 44–46, 116–117; and see A. Schalit, ''Die Denkschrift der Samaritaner an König Antiochos Epiphanes zu Beginn der grossen Verfolgung der jüdischen Religion im Jahre 167 v. Chr.,'' *ASTI*, 8 (1972), pp. 131–183, and especially note 40.

59 TB Soṭah 22a; and cf. TB Pesaḥim 51a. A separate article should be devoted to the term בור which occurs here. A connection between the concepts בור and ‘am ha-aretz occurs also in the famous statement of Hillel [R. Hillel ?], which we have discussed previously: אין בור ירא חטא, ולא עם הארץ חסיד ''An uncultured person [בור] is not sin-fearing, nor is an ‘am ha-aretz pious'' (Avot ii, 5). Cf. also הדר... בחורים יראת חטא. הדר עם הארץ מארה. קרא ולא שנה הרי זה בור. שנה ולא קרא עם הארץ. קרא ושנה ולא פירש חכם. קרא ושנה ופירש נבון. לא קרא ולא שנה נוח לו שלא

In the Babylonian Talmud we find the surprising concepts of a Cuthaean ḥaver and a Cuthaean 'am ha-aretz. הכותי מזמנין עליו: אמאי, לא יהא אלא עם הארץ, ותניא, אין מזמנין על עם הארץ. אביי אמר, בכותי חבר. רבא אמר, אפילו תימא בכותי עם הארץ, והכא בעם הארץ דרבנן דפליגי עליה דרבי מאיר עסקינן, דתניא, איזהו עם הארץ וכו׳ "A Cuthaean may be reckoned in for a zimmun.a) Why is this so? Wherein is he better than an 'am ha-aretz, and it has been taught, An 'am ha-aretz is not reckoned in for a zimmun? Abbaye answered, It refers to a Cuthaean who is a ḥaver. Rava said, You may even take it to refer to a Cuthaean who is an 'am ha-aretz, the reference here being to an 'am ha-aretz as defined by the Rabbis, who in this matter join issue with R. Meir. For it has been taught, Who is an 'am ha-aretz?..."[60]

These concepts obviously came into being a considerable time after the historical situation in which the concepts and actual appearance of the ḥaverim and the 'ammei ha-aretz occurred. We have previously mentioned that these concepts, which were connected with Eretz Israel, ceased to have any relevance at the beginning of the amoraic period, so that their presence in this passage is theoretical and devoid of any real, historical significance.[61] The same applies to the distinction between עם הארץ דרבי מאיר "the 'am ha-aretz as defined by R. Meir," and

נברא "...The glory of young men is the fear of sin. The glory of an 'am ha-aretz is destruction. One who has learnt Scripture but not Mishnah is an uncultured person [בור]. One who has learnt Mishnah but not Scripture is an 'am ha-aretz. One who has learnt Scripture and Mishnah but cannot explain them is a Sage. One who has learnt Scripture and Mishnah and is able to explain them is a man of intelligence. One who has learnt neither Scripture nor Mishnah were better not born" (Derekh Eretz Zuṭa x, Tosefta Derekh Eretz i, 8, ed. Higger, p. 247); etc. The grading, such as occurs in the minor tractates, is to a large extent clearly artificial and does not reflect the class and social stratification which is to be derived from the historical situation of this or that period.

60 TB Berakhot 47b; and see Diḳduḳei Soferim. Cf. also TB Niddah 33b: ורמינהי: על שישה ספקות שורפין את התרומה, על ספק בגדי עם הארץ... הכא במאי עסקינן בכותי חבר וכו׳ "But have we not learnt to the contrary: In six doubtful cases of impurity terumah is burnt [and one of them is] the doubtful impurity of the clothes of an 'am ha-aretz... Here we are dealing with the case of a Samaritan who was a ḥaver..."

61 In the parallel discussion in the Jerusalem Talmud (TJ Berakhot vii, 11b) these concepts are not mentioned.

a) When three or more men partake of a common meal, one of them, who conducts the grace after the meal, commences by inviting the others to join in the recital of the grace, and to each part of the introductory ritual recited by him, they make responses. This ceremony of inviting those present to participate in the saying of the grace after the meal is called zimmun.

עם הארץ דרבנן "the ‘am ha-aretz as defined by the Rabbis." These different definitions of the concept of the ‘am ha-aretz obviously do not refer to two different social strata, so that from the point of view of historical reality it is meaningless to make a distinction, which occurs several times in discussions in the Babylonian Talmud, between the ‘ammei ha-aretz as defined by R. Meir and the ‘ammei ha-aretz as defined by the Sages. It is also doubtful whether the concept of Cuthaeans in this passage is founded on a knowledge of the historical-social situation of the Samaritans or whether it is merely mentioned in the course of a theoretical discussion. There is evidence that Abbaye, one of the disputants in this discussion, knew Cuthaeans in Babylonia.[62] The theoretical possibility of a Cuthaean ḥaver, mentioned in the Babylonian Ṭalmud, may be based on the fact that the Cuthaeans in the diaspora were closer to the Jews than were those in Eretz Israel.

To sum up: there is, of course, no valid reason whatsoever for identifying the ‘ammei ha-aretz with the Samaritans. At the same time, there are in several areas parallel halakhot relating to both the Samaritans and the ‘ammei ha-aretz. The ‘am ha-aretz le-mitzvot was suspected in particular of not observing the commandments of the tithes, the Sabbatical Year, and purity, and so too, as we have found, were the Samaritans in part and not in such detail. The ‘am ha-aretz la-Torah was condemned for not engaging in the study of the Torah and for his hatred of the Sages, and it is clear that the Samaritans were similarly condemned. But whereas the problem of close social relations or of social shunning was naturally greater in the case of the Samaritans, whose status was that between a Jew and a gentile, such a question never arose in connection with the ‘ammei ha-aretz, who were always regarded as Jews. This difference found expression, in particular from the Bar Kokheva revolt onwards, in a growing political tension between Jews and Samaritans.

---

[62] TB Giṭṭin 45a.

BIBLIOGRAPHY

ALBECK, CH., *Das Buch der Jubiläen und die Halacha* (Bericht der Hochschule für die Wissenschaft des Judentums), Berlin, 1930.

ALLON, G., *Meḥkarim be-Toledot Yisra'el*, I–II, Tel Aviv, 1957–1958.

ALLON, G., *Toledot ha-Yehudim be-Eretz Yisra'el bi-Teḳufat ha-Mishnah ve-ha-Talmud*, I–II, Tel Aviv, I: 1958[3], II: 1961[2].

AMUSSIN, I. D., "עם הארץ," *Vestnik Drevney Istorii*, 2 (1955), pp. 14–36.

AVI-YONA, M., *Bi-Yemei Roma u-Byzantiyon*, Jerusalem, 1970[4].

BACHER, W., "Zur Geschichte der Schulen Palästina's im 3. und 4. Jahrhundert: Die Genossen (חברייא)," *MGWJ*, 43 (1899), pp. 345–360.

BAER, Y., "The Historical Foundations of the Halakhah" (Hebrew), *Zion*, 17 (1952), pp. 1–55.

BARON, S., *A Social and Religious History of the Jews*, I–II, New York, 1952[2].

BELKIN, S., *Philo and the Oral Law*, Cambridge (Mass.), 1940.

BENOIT, P., O.P., J. T. MILIK, R. DE VAUX, O.P., *Les Grottes de Murabba'at* (*Discoveries in the Judaean Desert*, II), Oxford, 1961.

BERGMANN, J., "Schebua Ha-ben," *MGWJ*, 76 (1932), pp. 465–470.

BICKERMANN, E., "Zur Datierung des Pseudo-Aristeas," *ZNW*, 29 (1930), pp. 280–298.

BLAU, L., *Papyri und Talmud in gegenseitiger Beleuchtung*, Leipzig, 1913.

BOUSSET, W., *Die Religion des Judentums im späthellenistischen Zeitalter*, 3. verb. Aufl. hrsg. von H. Gressmann, Tübingen, 1926 (1st ed. Berlin, 1903).

BRÜLL, N., "The Basis and Development of the Laws relating to the Purity of the Hands" (Hebrew), *Bet Talmud*, II (1881), pp. 315–320, 325–333, 368–374; III (1882), pp. 23–26, 49–52.

BÜCHLER, A., *Die Priester und der Cultus im letzten Jahrzehnt des jerusalemischen Tempels* (Jahresbericht der isr. theol. Lehranstalt), Vienna, 1895.

BÜCHLER, A., "Die priesterlichen Zehnten und die römischen Steuern in den Erlässen Caesars," *Steinschneider-Festschrift*, Leipzig, 1896, pp. 91–109.

BÜCHLER, A., *Der galiläische 'Am-ha'Areṣ des zweiten Jahrhunderts*, Vienna, 1906.

BÜCHLER, A., "La pureté lévitique de Jérusalem et les tombeaux des prophètes," *REJ*, 62 (1911), pp. 201–215; 63 (1912), pp. 30–50.

BÜCHLER, A., "Learning and Teaching in the Open Air in Palestine," *JQR* (N.S.), 4 (1913–1914), pp. 485–491.

BÜCHLER, A., "The Levitical Impurity of the Gentile in Palestine before the Year 70," *JQR* (N.S.), 17 (1926), pp. 1–81.

BÜCHLER, A., "He'arot ve-He'arot 'al Matzav ha-Ishah be-Sefer Yehudit," *Sefer Blau*, Budapest, 1926, pp. 42–67.

BÜCHLER, A., "Die Schammaiten und die levitische Reinheit des עם הארץ," *Freimann-Festschrift*, Berlin, 1937, pp. 21–37.

CANTINEAU, J., *Le Nabateén*, I–II, Paris, 1930–1932.

CARL, Z., "The Tithe and the Terumah" (Hebrew), *Tarbiz*, 16 (1945), pp. 11–17.

CHURGIN, P., "Sefer Yehudit," *Meḥkarim bi-Teḳufat Bayit Sheni*, New York, 1949, pp. 123–147.

CHWOLSON, D., "Wer und was ist Am-Haarez עם הארץ in der alten rabbinischen Literatur?" *Beiträge zur Entwicklungsgeschichte des Judentums*, Leipzig, 1910, pp. 1–54.

COWLEY, A., *The Samaritan Liturgy*, I–II, Oxford, 1909.

DAICHES, S., "The Meaning of עם הארץ in the Old Testament," *JThS*, 30 (1929), pp. 245–249.

DALMAN, G., *Orte und Wege Jesu*, Gütersloh, 1924[3].

DE VAUX, R., *Les institutions de l'Ancien Testament*, I–II, Paris, 1958–1960.

DE VAUX, R., "Le sens de l'expression 'Peuple du pays' dans l'Ancien Testament et le rôle politique du Peuple en Israel," *RA*, 58 (1964), pp. 167–172.

DÖLLER, J., *Die Reinheits- und Speisegesetze des Alten Testaments*, Münster, 1917.

EPSTEIN, J. N., *Mavo le-Nusaḥ ha-Mishnah*, I–II, Jerusalem, 1964[2].

EPSTEIN, J. N., *Mevo'ot le-Sifrut ha-Tanna'im*, Jerusalem, 1957.

EPSTEIN, J. N., *Mevo'ot le-Sifrut ha-Amora'im*, Jerusalem, 1962.

FINKELSTEIN, L., *Akiba*, New York, 1936.

FINKELSTEIN, L., *The Pharisees*, I–II, Philadelphia, 1962[3].

FLUSSER, D., "The Religious Ideas of the Judean Desert Sect" (Hebrew), *Zion*, 19 (1954), pp. 89–103.

FRANKEL, Z., *Darkhei ha-Mishnah*, Adapted and Revised by Yitzhak Nissenbaum, Warsaw, 1923 (1st ed. Leipzig, 1859).

FRIEDLÄNDER, M., *Zur Entstehungsgeschichte des Christentums*, Vienna, 1894, pp. 37–58.

FRIEDLÄNDER, M., *Die religiösen Bewegungen*, Berlin, 1905, pp. 78–88.

GASTER, M., *The Samaritan Literature*, London, 1925.

GEIGER, A., *Urschrift und Übersetzungen der Bibel*, Frankfort-on-Main, 1928.

GILAT, Y. D., "Mi-de-Orayeta li-de-Rabbanan," *Sefer Zikkaron le-Binyamin de-Vries*, Jerusalem, 1968, pp. 84–93.

GILLISCHEWSKI, E., "Der Ausdruck עם הארץ im A.T.," *ZAW*, 40 (1922), pp. 137–142.

GINZBERG, L., *Perushim ve-Ḥiddushim ba-Yerushalmi*, I–IV, New York, 1941–1961.

GINZBERG, L., *'Al Halakhah ve-Aggadah*, Tel Aviv, 1960.

GIPSEN, W. H., "Clean and Unclean," *OTS*, 5 (1948), pp. 190–196.

GRAETZ, H., *Geschichte der Juden*, Leipzig, 1906[5].

GRINSPAN, N. S., *Mishpaṭ 'Am ha-Aretz be-Sifrut ha-Halakhah be-Kol Teḳufoteha*, Jerusalem, 1946.

GRINTZ, Y. M., "Adat ha-Yaḥad — Issiyyim — Bet [I]sin," *Peraḳim be-Toledot Bayit Sheni*, Jerusalem, 1969, pp. 105–142.

GRODIS, R., "Sectional Rivalry in the Kingdom of Judah," *JQR* (N.S.), 25 (1934–1935), pp. 237–259.

HALEVY, I., *Dorot ha-Ri'shonim*, I–VI, Pressburg–Frankfort-on-Main, 1901–1918.

HARNACK, C., *Die Mission und Ausbreitung des Christentums*, Leipzig, 1924.

HEINEMANN, J., *Ha-Tefillah bi-Teḳufat ha-Tanna'im ve-ha-Amora'im*, Jerusalem, 1966[2].

HERFORD, R. TRAVERS, *The Pharisees*, London, 1924 (New York, 1924; Boston, 1962).

Herr, M. D., "Gezerot ha-Shemad ve-Ḳiddush ha-Shem bi-Yemei Hadrianus," *Milḥemet Ḳodesh u-Martyrologyah be-Toledot Yisra'el u-ve-Toledot ha-'Ammim*, Historical Society of Israel, Jerusalem, 1967, pp. 73–92.
Hershkovitz, Y., "The Cuthaeans in Statements of the Tannaim" (Hebrew), *Yavneh*, 2 (1940), pp. 71–105.
Herzfeld, L., *Geschichte des Volkes Jisrael*, II, Leipzig, 1863.
Higger, M., "Pirḳei Rabbenu ha-Ḳadosh," *Ḥorev*, 6 (1941), pp. 116–149.
Hoenig, S., "Historical Inquiries, I, Heber Ir," *JQR* (N.S.), 48 (1957–1958), pp. 123–139.
Horovitz, J., "חבר עיר," *Festschrift zum siebzigsten Geburtstag J. Guttmanns*, Leipzig, 1915, pp. 125–142.
Horovitz, J., "Nochmals חבר עיר. Bemerkungen zu des Herrn Prof. Krauss," *JJLG*, 17 (1926), pp. 241–315.
Ish-Shalom, M., "The History of the Uncleanness of Hands in General and through Touching the Sacred Scriptures in Particular" (Hebrew), *Ha-Goren*, 2 (1900), pp. 66–74; 3 (1902), pp. 30–39.
Jeremias, J., *Jerusalem zur Zeit Jesu*, Göttingen, 1962³.
Joël, M., *Blicke in die Religionsgeschichte*, I–II, Breslau, 1880–1883.
Katz, B-Z., *Perushim, Tzeduḳim, Ḳanna'im, Notzerim*, Tel Aviv, 1948, pp. 25–34.
Kaufmann, Y., *Toledot ha-Emunah ha-Yisra'elit*, I–VIII, Jerusalem and Tel Aviv, 1936–1956.
Klamroth, E., *Die jüdischen Exulanten in Babylonien*, Leipzig, 1912, pp. 99–101.
Klausner, J., *Hisṭoryah shel ha-Bayit ha-Sheni*, I–V, Jerusalem, 1958⁵.
Klausner, J., *Jesus of Nazareth*, London and New York, 1925.
Klein, S., "On the History of the Great Land Tenancy in Eretz Israel" (Hebrew), *IEJ*, I, 3 (1933), pp. 3–9.
Klein, S., *Sefer ha-Yishuv*, I, Jerusalem, 1939.
Klein, S., *Eretz Yehudah*, Tel Aviv, 1939.
Klein, S., *Eretz ha-Galil*, Jerusalem, 1967².
Koehler, L., "Aussatz," *ZAW*, 67 (1955), pp. 290–291.
Krauss, S., "חבר עיר. Ein Kapitel aus altjüdischer Kommunalverfassung," *JJLG*, 17 (1926), pp. 195–241.
Krochmal, N., *Moreh Nevukhei ha-Zeman*, Berlin, 1924.
Leszynsky, R., *Die Sadduzäer*, Berlin, 1912.
Lévy, I., *La légende de Pythagore de Grèce en Palestine*, Paris, 1927, pp. 236–263.
Lewin, B. M., *Otzar ha-Ge'onim*, I–XII, Haifa–Jerusalem, 1928–1943.
Lewy, I., *Über einige Fragmente aus der Mischna des Abba Saul*, Berlin, 1876 [translated into Hebrew in *Mesillot le-Torat ha-Tanna'im*, Tel Aviv, 1928, pp. 92–133].
Lieberman, S., "Emendations on the Jerushalmi" (Hebrew), *Tarbiz*, 3 (1932), pp. 210–212.
Lieberman, S., "Further Notes on the Leiden MS. of the Jerushalmi" (Hebrew), *Tarbiz*, 20 (1949), pp. 107–117.
Lieberman, S., "The Discipline of the So-Called Dead Sea Manual of Discipline," *JBL*, 71 (1952), pp. 199–206.
Lieberman, S., *Greek in Jewish Palestine*, New York, 1942.
Lieberman, S., *Hellenism in Jewish Palestine*, New York, 1950.
Lietzmann, L., *Geschichte der Alten Kirche*, Berlin, 1964.

MALAMAT, A., "The Last Kings of Judah and the Fall of Jerusalem," *IEJ*, 18 (1968), pp. 140–156.

MANTEL, H., *Studies in the History of the Sanhedrin*, Cambridge (Mass.), 1961.

MANTEL, H., "The Nature of the Great Synagogue (Knesset ha-Gedolah)," *Proceedings of the Fourth World Congress of Jewish Studies*, I, Jerusalem, 1967, pp. 81–88 (Hebrew), p. 258 (English Summary).

MEYER, E., *Ursprung und Anfänge des Christentums*, I–III, Stuttgart–Berlin, 1921–1923.

MONTGOMERY, J., "The Etymology of דמאי," *JQR* (N.S.), 23 (1932–1933), p. 209.

MOORE, G. F., *Judaism in the First Centuries of the Christian Era*, Cambridge (Mass.), 1927–1930.

NEUSNER, J., "The Fellowship (חבורה) in the Second Jewish Commonwealth," *HTR*, 53 (1960), pp. 125–142.

NICHOLSON, E. W., "The Meaning of the Expression עם הארץ in the Old Testament," *JSS*, 10 (1965), pp. 59–66.

OPPENHEIMER, A., "Hafrashat Ma'aser Ri'shon Halakhah le-Ma'aseh bi-Teḳufat Bayit Sheni," *Sefer Zikkaron le-Binyamin de-Vries*, Jerusalem, 1968, pp. 70–83.

PINELES, H. M., *Darkah shel Torah*, Vienna, 1861.

POLAND, F., *Geschichte des griechischen Vereinswesens*, Leipzig, 1909.

RABIN, CH., *Ha-Yaḥad, ha-Ḥavurah ve-ha-Issiyyim*, *'Iyyunim bi-Megillot Midbar Yehudah*, Jerusalem, 1957, pp. 104–122.

RABIN, CH., *Qumran Studies*, Oxford, 1957.

REYMOND, P., "L'eau, sa vie et sa signification dans l'Ancien Testament," Supplement VI to *VT*, 1958.

ROST, L., "Die Bezeichnungen für Land und Volk im Alten Testament," *Festschrift-Procksch*, Leipzig, 1934, pp. 125–148.

SAFRAI, S., "Siḳariḳon," *Zion*, 17 (1952), pp. 56–64.

SAFRAI, S., "The Holy Congregation in Jerusalem" (Hebrew), *Zion*, 22 (1957), pp. 183–193.

SAFRAI, S., "Bet She'arim in Talmudic Literature" (Hebrew), *Eretz-Israel*, 5 (1958), pp. 206–212.

SAFRAI, S., *Ha-'Aliyah le-Regel bi-Yemei Bayit Sheni*, Tel Aviv, 1965.

SAFRAI, S., "Teaching of Pietists in Mishnaic Literature," *JJS*, 16 (1965), pp. 15–33.

SAFRAI, S., "Sabbatical Year Commandments under the Conditions Prevailing after the Destruction of the Second Temple" (Hebrew), *Tarbiz*, 35 (1966), pp. 304–328; 36 (1967), pp. 26–46.

SAFRAI, S., "Ha-'Ir ha-Yehudit be-Eretz Yisra'el bi-Teḳufat ha-Mishnah ve-ha-Talmud," *Ha-'Ir ve-ha-Ḳehillah*, Historical Society of Israel, Jerusalem, 1967, pp. 227–236.

SAFRAI, S., "Elementary Education, Its Religious and Social Significance in the Talmudic Period," *Cahiers d'Histoire Mondiale*, 11 (1968), pp. 148–169.

SAFRAI, S., "Beḥinot Ḥadashot li-Be'ayat Ma'amado u-Ma'asav shel Rabban Yoḥanan ben Zakkai le-aḥar ha-Ḥurban," *Sefer Zikkaron le-Allon*, Tel Aviv, 1970, pp. 203–226.

SAFRAI, S., *R. 'Aḳiva ben Yosef — Ḥayyav u-Mishnato*, Dorot Library, Jerusalem, 1970.

SAFRAI, S., "The Holy Congregation in Jerusalem," *Scripta Hierosolymitana*, XXIII, Studies in History, 1972, pp. 62–78.

SCHALIT, A., *Ha-Mishṭar ha-Roma'i be-Eretz Yisra'el*, Jerusalem, 1937.

SCHALIT, A., *Namenwörterbuch zu Flavius Josephus*, Leyden, 1968.

SCHALIT, A., *König Herodes, der Mann und sein Werk*, Berlin, 1969.

SCHALIT, A., "Die Denkschrift der Samaritaner an König Antiochos Epiphanes zu Beginn der grossen Verfolgung der jüdischen Religion im Jahre 167 v. Chr.," *ASTI*, 8 (1972), pp. 131–183.

SCHEFTELOWITZ, J., "Das Opfer der roten Kuh (Num. 19)," *ZAW*, 39 (1921), pp. 113–123.

SCHÜRER, E., *Geschichte des jüdischen Volkes im Zeitalter Jesu Christi*, Leipzig, 1907[4].

SCHWARZ, A., "Ein von Resch Lakisch angedeuteter Kopistenfehler in der Mischna," *MGWJ*, 71 (1927), pp. 8–13.

SEELIGMANN, I., *The Septuagint Version of Isaiah*, Leyden, 1948.

SLOUSCH, N., "Representative Government among the Hebrews and Phoenicians," *JQR* (N.S.), 4 (1913–1914), pp. 303–310.

SOGGIN, J. A., "Der judäische 'am-ha'areṣ und das Königtum in Juda," *VT*, 13 (1963), pp. 187–195.

STERN, M., *Ha-Te'udot le-Mered ha-Ḥashmona'im*, Tel Aviv, 1965.

SULZBERGER, M., *The Am ha-aretz, The Ancient Hebrew Parliament*, Philadelphia, 1910[2].

TADMOR, H., " 'The People' and the Kingship in Ancient Israel," *Journal of World History*, 11 (1968), pp. 56–68.

TALMON, S., "The History of the 'Am ha-Aretz in the Kingdom of Judah" (Hebrew), *Bet Miḳra*, 31 (1967), pp. 27–55.

TALMON, S., "The Judaean 'am-ha'areṣ in the Historical Perspective," *Proceedings of the Fourth World Congress of Jewish Studies, 1965*, Jerusalem, 1967, I, pp. 71–76.

TCHERIKOVER, V., *Ha-Yehudim be-Mitzrayim ba-Teḳufah ha-Hellenisṭit Romit le-Or ha-Papyrologyah*, Jerusalem, 1963[3].

TCHERNOWITZ, CH. (Rav Tzair), *Toledot ha-Halakhah*, I–IV, New York, 1935–1950.

TCHERNOWITZ, CH. (Rav Tzair), "Demai — Hatza'ah Hisṭorit le-Taḳḳanot Yoḥanan Kohen Gadol she-ba-Mishnah," *Jewish Studies in Memory of G. A. Kohut*, New York, 1935, pp. 46–58.

URBACH, E., " 'Am ha-Aretz," *Proceedings of the World Congress of Jewish Studies, 1947*, Jerusalem, 1952, pp. 362–366.

URBACH, E. E., "Social and Religious Trends in the Sages' Doctrine of Charity" (Hebrew), *Zion*, 16 (1951), pp. 1–27.

URBACH, E. E., "Ascesis ve-Yissurim be-Torat Ḥazal," *Sefer ha-Yovel li-Kevod Yitzḥaḳ Baer*, Jerusalem, 1960, pp. 48–68.

URBACH, E. E., "Ma'amad ve-Hanhagah be-'Olamam shel Ḥakhmei Eretz Yisra'el," *Divrei hε Aḳademyah ha-Le'ummit ha-Yisra'elit le-Madda'im*, II, 4, Jerusalem, 1969, pp. 31–54.

URBACH, E. E., *Ḥazal — Pirḳei Emunot ve-De'ot*, Jerusalem, 1969. In the meantime the work has appeared in English: *The Sages, Their Concepts and Beliefs*, Jerusalem, 1975.

VOGELSTEIN, H., *Der Kampf zwischen Priestern und Leviten*, Stettin, 1889.

VON RAD, G., *Deuteronomiumstudien*, Göttingen, 1948[2].

WÄCHTER, TH., "Reinheitsvorschriften im griechischen Kult," *Religionsgeschichtliche Versuche und Vorarbeiten*, IX, 1, Giessen, 1910.

WALLACE, S. L., *Taxation in Egypt from Augustus to Diocletian*, Princeton, 1938.

WEBER, M., *Das antike Judentum*, Tübingen, 1921.

WEINBERG, M., "Die Organisation der jüdischen Ortsgemeinden in der talmudischen Zeit," *MGWJ*, 41 (1897), pp. 588–604, 639–660, 673–691.

WEISS, I. H., *Dor Dor ve-Doreshav*, I–V, Vilna, 1904[4].

WELLHAUSEN, J., *Prolegomena zur Geschichte Israels*, Berlin, 1905[6].

WELLHAUSEN, J., *Israelitische und jüdische Geschichte*, Berlin, 1958[9].

WOLFF, C. U., "Traces of Primitive Democracy in Ancient Israel," *JNES*, 6 (1947), pp. 98–108.

WÜRTHWEIN, E., *Der 'amm ha'arez im AT*, Stuttgart, 1936.

ZEITLIN, S., "The Am haarez," *JQR* (N.S.), 23 (1932), pp. 45–61.

ZEITLIN, S., "Johanan the High Priest's Abrogations and Decrees," *Studies and Essays in Honour of A. A. Neuman*, Leyden, 1962.

ZEMIRIN, S., *Yo'shiyyahu u-Teḳufato*, Jerusalem, 1952.

ZIEBARTH, E., *Das griechische Vereinswesen*, Leipzig, 1896.

INDEX OF NAMES

Names, partly enclosed in square brackets, occur in both the full and the shorter forms

INDEX OF SOURCES

SCRIPTURAL REFERENCES

RABBINIC SOURCES

SAMARITAN AND KARAITE SOURCES